Structure, Agency and Biotechnology

Structure, Agency and Biotechnology

The Case of the Rothamsted GM Wheat Trials

Aristeidis Panagiotou

ANTHEM PRESS

Anthem Press
An imprint of Wimbledon Publishing Company
www.anthempress.com

This edition first published in UK and USA 2019
by ANTHEM PRESS
75–76 Blackfriars Road, London SE1 8HA, UK
or PO Box 9779, London SW19 7ZG, UK
and
244 Madison Ave #116, New York, NY 10016, USA

First published in the UK and USA by Anthem Press 2017

British Library Cataloguing-in-Publication Data
A catalogue record for this book is available from the British Library.

ISBN-13: 978-1-78527-124-3 (Pbk)
ISBN-10: 1-78527-124-5 (Pbk)

This title is also available as an e-book.

For Dimitra

CONTENTS

ILLUSTRATIONS

Figures

Tables

ABBREVIATIONS

ABC	Agricultural Biotechnology Council
ACRE	Advisory Committee on Releases to the Environment
ANT	Actor-Network Theory
BBSRC	Biotechnology and Biological Sciences Research Council
BIS	Department for Business Innovation & Skills (UK Government)
CSA	Chief Scientific Adviser
CST	Council for Science and Technology
DEFRA	Department for Environment, Food & Rural Affairs
EC	European Commission
EFSA	European Food Safety Authority
EPOR	Empirical Program of Relativism
EU	European Union
GMO(s)	Genetically Modified Organism(s)
ILSI	International Life Science Institute
IRGC	International Risk Governance Council
MEP	Member of the European Parliament
SCOT	Social Construction of Technology
ST	Structuration Theory
SSK	Sociology of Scientific Knowledge
SST	Strong Structuration Theory
STC	Science and Technology Committee
STS	Science and Technology Studies
TAI	Technology, Appropriation, Ideology Scheme

ACKNOWLEDGMENTS

This book is a revised and considerably enriched version of my PhD thesis. As such, I would like to thank all the scholars who were involved in one way or another with this challenging and deeply gratifying process. Ted Benton, for guiding me through the rough seas of academic writing with profound dedication and calmness. Brian Wynne, for showing a keen interest in my writings and encouraging me to flesh out my arguments with greater confidence. Lydia Morris, for being equally supportive and for showing me how details can play a decisive role in the overall thrust of an argument. Bryan S. Turner, for being very sympathetic to this publishing endeavor from the very beginning, and for making me think of the whole GM problematic in more pragmatistic terms. Les Levidow, for the very thought-provoking and warm discussions we had across the various European capitals we encountered each other during the past couple of years. As a student, I was inspired by their writings, but it was meeting them in person that made all the difference.

I would especially like to thank Rob Stones for all the academic, mental and moral support he has so generously offered all these years. He has acted as a father figure to me, and I wouldn't be writing these lines if not for his encouragement in the first place.

I would also like to thank Michael Biggs at Oxford University for teaching me how to be specific in my writings and ambitious in my goals. My deepest gratitude goes to the Sociology academics of the American College of Greece who instilled in me the principles of the discipline with remarkable clarity and ethos: Spyros Gangas, Tina Katsarou, Gregory Katsas, Georgia Lagoumitzi and Iordanis Psimmenos.

Finally, but by no means least, I would like to thank everyone at Anthem Press involved with this project, from proposal reviewing to typesetting. In particular, I would like to acknowledge the contribution of the two anonymous reviewers for helping make the overall organization of this book more efficient and the articulation of some arguments more effective.

This book is dedicated to my wife, Dimitra, for so graciously offering to put aside her own dreams and aspirations so that I could chase mine …

Chapter 1

A HOLISTIC APPROACH TO THE GM CONTROVERSY

The GM Controversy as a "Lightning Rod"

Over the past 20 years, biotechnology has gradually shifted from a paradigm of purely scientific research and experimentation to worldwide commercialization in a variety of industries, from agriculture and food production to chemicals and pharmaceuticals. While the rapidly expanding number of biotechnological applications and products has been met with skepticism or even fear, among consumers and public authorities, some societies, especially in North America, have proved to be more willing to accept the new technologies. In Britain and Europe, in general, genetically modified (GM) crops and food have become a *cause célèbre* among environmentalists and consumer protesters (Falkner 2000, 300). On many occasions, GM food—also frequently called by activists as "GM pollution" or *Gen-Müll* ("genetic garbage")—has been portrayed as an impurity contaminating science, agriculture, the environment and even democratic sovereignty. When the first US shipments containing GM soya and maize arrived at European ports in the fall of 1996, a concerted Europe-wide symbolic protest generated publicity around the fact that GM grain was entering processed food without labeling. Governments and agricultural biotechnology (agbiotech) companies were accused of "force-feeding [consumers] GM food" and, as a response, NGOs carried out surveillance of food products for "GM contamination." Since then, the protest against genetically modified organisms (GMOs) has ranged from symbolic moves with quasi-theatrical elements to resolute "decontamination" actions with potentially serious legal repercussions.[1] Regarded as a technology with effects of potentially apocalyptic dimensions, the phrase "GM-free" quite often has had connotations similar to those of "nuclear-free" in the 1980s (Levidow 2009, 110).

Protests against GMOs have, however, never been strictly targeted toward GMOs per se; rather, they have always encompassed issues that stretch well beyond the particular locus of discontent. The anxieties about the safety of GMOs are certainly not the only issues in the GM controversy. Such concerns are "the entry point" for understanding what is at stake with GM technology and should be seen as "the start of the discussion rather than the end" (Sciencewise 2011, 3). Other considerations that are often part and parcel of the GM discussion include the issue of sustainability and the environmental impacts of GM technologies; questions about intellectual property, patenting, the livelihoods of developing country farmers; and the questions of democratic governance and sound regulation (Sciencewise 2011, 3, 4). Therefore, instead of framing the commercialization of GM technology as a binary decision, it would be more consistent with the messages articulated by interested parties to envisage the conjuncture as "a lightning rod for many other issues—about fairness, access and corporate control of the food system" (Burrows, qtd. in House of Commons Science and Technology Committee 2014a, 11).

Public Sentiment, Scientific Viewpoints and Legislative Frameworks: A Dissonant Coexistence

The EU regulatory system

The concern about agbiotech has been vibrant not only across various sectors of the civil society but has also stirred a controversy among authorities at an international level. While in the United States GM foods are placed on the market without being subject to any form of mandatory labeling (World Health Organization 2005, 51), in the EU, legislation is much stricter. At the time of writing, for a GM crop to be imported, tested, cultivated or marketed in a European country, very firm criteria should be adhered to (European Commission, 2012d). However, the EU did not have such a strict regulatory system from the very beginning. In fact, a number of GM crops were approved in the mid-1990s (including the Flavr Savr GM tomato paste that was sold and subsequently withdrawn in the UK in 1996), but as the issue of GM technology became increasingly controversial, the EU faced intense pressure from its member states to develop a more robust regulatory system. As a consequence, a de facto moratorium was placed by the EU on GM products, starting October 1998 (The European Union Center of North Carolina 2007, 1). While member states debated and eventually passed new legislation on the approval of GMOs and their labeling and traceability standards, in 2003 the United States filed a complaint with the World Trade Organization (WTO) arguing that the moratorium the EU and six member states had maintained was illegal, costing US exports $300 million per year (World Trade Organization 2008). The European Commission (EC) characterized the filing of the complaint as "legally unwarranted, economically unfounded and politically unhelpful." The EU Commissioner for the Environment also added: "This US move is unhelpful. It can only make an already difficult debate in Europe more difficult. [...] We should not be deflected or distracted from pursuing the right policy for the EU" (European Commission 2003a). With the complaint at the WTO pending, the EU adopted Regulation 1829/2003 on GM food and feed and Regulation 1830/2003 on GMOs, traceability, labeling and derived food and feed (European Commission 2012b). Although in 2006 the WTO ruled against the EU, the United States, supported by Canada and Argentina, was still unhappy with the procedures that it found too convoluted and "based on political expediency more than on health or safety concerns" (Euractiv 2006).

Partially as a response to international discontent, in mid-2010 the Commission proposed a new set of rules for the authorization of GMOs that would allow EU countries to restrict or ban GMO cultivation on their territory by using any acceptable reason under the treaty establishing the European Community without undermining the EU risk assessment, which would remain unchanged (European Commission 2012c). In the same year, the European Commission approved the cultivation of the GM potato "Amflora" developed by BASF—the first GM cultivation approval in 12 years (Euractiv 2010) and, one year later, the EU allowed traces of unapproved GM material in animal feed imports (Euractiv 2011). The year 2011 was also when EC president Manuel Barroso appointed the Commission's first chief scientific adviser (CSA), Professor Anne Glover (Europa 2011). Professor Glover made her affinity to GMOs well known in a number of interviews she gave by overtly expressing her disapproval of the *precautionary principle*, the fundamental notion all EU directives on GMOs abide by (Euractiv 2012c), and by her willingness to eat GM food if it were approved in Europe (EuropaBio 2012d). The precautionary principle—which asks scientists and policymakers under circumstances of uncertainty to err on the side of caution—is a fundamental notion in the GM debate and will be extensively discussed in Chapter 7. The contribution of Anne Glover as CSA was, however, short-lived. The newly elected president of the EC, Jean-Claude Juncker, swiftly removed Glover from the position of CSA by reaffirming, nonetheless, his commitment to "independent scientific advice" (Fleming 2014). President Juncker's decision to relieve Glover of her duties and eventually axe the position of CSA was met with opposing reactions. NGOs such as Greenpeace and Corporate Europe Observatory endorsed the decision, as they discerned "fundamental flaws" in the role of the CSA, which made "the influence of corporate lobbyists [...] even easier" (Nelsen, "NGO backlash to Chief Scientific Advisor position grows," 2014). At the same time, Conservative Members of the European Parliament (MEPs), the European Association for Bio-industries (EuropaBio) and certain scientific circles reacted angrily to the news of Glover's departure and accused President Juncker of caving in to the Green lobby (Fleming 2014; Delingpole 2014).

In 2014, after almost a decade of legal battles, the EU policy on GMOs changed direction once again. In June of that year, the EU approved the 2010 EC proposal and finally reached an agreement to allow its member states to restrict or ban cultivation of GMO crops on their territory by adopting an opt-out measure. By October 3, 2015, 19 of the 28 EU member states (including Austria, Denmark, France, Germany, Greece, Italy, the Netherlands and Poland) informed the EC that they wished to opt out of new GMO cultivation approvals. Belgium and Britain, on the other hand, asked that the opt-out mechanism be applicable to only certain parts of their countries' territories (DW 2015). This move has been considered as a considerable blow to the biotech industry, as it is estimated that with the new EU rules, around two-thirds of the EU's population and arable land will be GM-free; that is, only 140,000 hectares of land will be cultivated with GM crops in the EU, compared to 181m hectares in the rest of the world (Nelsen 2015).

England decides to endorse GM cultivation while the rest of the UK opt out

The divided opinion on GMOs within the EU is also apparent among UK countries. While Northern Ireland, Scotland and Wales have all adopted the opt-out rules, England has decided to allow GM crop cultivation. The British government's endorsement of GM technology and general affinity to the use of advanced genetic techniques in crop improvement has been overtly expressed by its members. In an often-quoted speech to the National Farmers' Union in February 2013, Owen Patterson, MP, the secretary of state for the Department for Environment, Food & Rural Affairs (DEFRA), voiced his concern that "the rest of the world is ploughing ahead and reaping the benefits of this technology while Europe risks being left behind." The EU regulatory arrangements were construed as being in a state of "paralysis" as the UK government was looking at "how best to capitalize on the UK's world-class science and technology [...] and take advantage of opportunities to export UK agri-tech skills and services" (Patterson 2013). The EU approval process was also the point of criticism by DEFRA minister, Lord de Mauley, who characterized it as an "unduly slow operation [...] that is deterring investment and innovation in this technology" (Case 2012a). The commitment to reaping the potential benefits of GM technology was reaffirmed by Patterson's replacement, Elizabeth Truss, MP, who argued that "[GM crops] have a role to play here in Britain" and that "[British] farmers need access to the technology that will help them work in world markets" (Webster 2015). The optimism that GM technology heralds, however, is not shared by the governments of the other UK nations. As GM policy is devolved within the UK, the policies of Northern Ireland, Scotland and Wales, which espoused the opt-out clause of the EU, come in direct contrast to that of the UK government. Richard Lochhead, Scotland's environment secretary, said he wanted to uphold the precautionary principle since he believed that the potential risks to other crops and wildlife from GMOs outweighed the likely benefits of the technology. He also clearly expressed his long-standing concerns about GM crops and argued that "allowing GM crops to be grown in Scotland would damage our clean and green brand, thereby gambling with the future of our £14bn food and drink sector" (The Scottish Government 2015). The decisions of Northern Ireland and Wales to ban GM crops were articulated around similar discursive motifs. Environment Minister Mark Durkan announced that he remained "unconvinced of the advantages of GM crops" and considered it "prudent to prohibit their cultivation here for the foreseeable future." He also added that "we are rightly proud of our natural environment and rich biodiversity[…] I am concerned that the growing of GM crops, which I acknowledge is controversial, could potentially damage that image" (Northern Ireland Executive 2015). The Welsh Deputy Minister for Farming and Food, Rebecca Evans, for her part, also stressed the "need to preserve consumer confidence and maintain our focus on a clean, green, natural environment. By having the ability to control what is grown in Wales, we can have confidence in preserving these values" (qtd. in Sarich 2015).

The STC inquiry

The British government's decision to allow cultivation of GM crops in England was not simply a political choice, but was a decision endorsed by numerous independent scientists, scientific bodies, research centers and professional associations with or without declared interests in the technology. On February 14, 2014—that is, almost twenty months before the government's final decision—the House of Commons Science and Technology Select Committee (STC), whose role is to "ensure that Government policy and decision-making are based on good scientific and engineering advice and evidence" (UK Parliament 2015), launched an inquiry on "GM foods and application of the precautionary principle in Europe." The chair of the STC, Andrew Miller, MP, explained that "GM technology potentially offers an array of benefits, but concerns are being expressed that it is being held back by misuse of the precautionary principle" and that the purpose of the inquiry was to assess "whether such restrictions are hampering UK scientific competitiveness, and whether they are still appropriate in light of the available evidence on the safety of GM" (qtd. in Committee 2014). By April 23, more than sixty influential individual and collective actors with specialties in the field responded to the STC's call for written evidence on the effectiveness of EU and UK safety regulations, the existence of barriers to the conduct of research on GM foods and the appropriateness of the application of the precautionary principle in the EU and the UK (Committee 2014). In October 2014, contributions to the cross-parliamentary inquiry launched by the STC were further expanded with the invitation of more than thirty interested members to provide oral evidence in the House of Commons on the regulations restricting the growth of GM foods in the UK and across the European continent.

The submitted evidence highlighted the profoundly controversial nature of the GM debate. One thing that became apparent from the STC inquiry is that there is not only stark disagreement on the appropriateness of the precautionary principle as a valid tool for risk assessment, but there is also a lack of consensus on fundamental issues intrinsic to GM technology. Four overarching themes can be discerned across apparent advocates[2] of GMOs:

- There is *no* credible scientific evidence against the safety of GMOs to human health and the environment;
- There are significant benefits in GMOs as these can play a major role in meeting the future challenges of global population growth, climate change effects and food security;
- The precautionary principle is either irrelevant or misused;
- The EU regulation is sluggish and heavily politicized.

The cluster of individual and collective actors who developed their positions along these four main blocks include, among others, advisory bodies such as the Advisory Committee on Releases to the Environment (ACRE), political organizations such as DEFRA and the Department for Business, Innovation & Skills (BIS), corporations and corporate groups such as the Agricultural Biotechnology Council (ABC), BASF plc, Bayer CropScience Ltd, independent scientific bodies and research centers such as the James Hutton Institute, the John Innes Centre, the Nuffield Council of Bioethics, Sense about Science and the Science Council, and also scientists who specialized in the field, such as Professor C. J. Pollock (chair of the Scientific Steering Committee, which oversaw the Farm-Scale Trials of GM herbicide-tolerant crops, and chair of ACRE for 14 years up to 2013), Sir Mark Walport (the CSA[3] of the UK government and cochair of the Council for Science and Technology (CST)), Dr. Julian Little (chair of ABC), Sir David Baulcombe (Regius Professor of Botany, University of Cambridge, and main author of the "GM science update" report submitted to CST, often referred to as the "Baulcombe report") and more.

The expressed certainty on the safety of GM foods and the optimism on the potential benefits of GM technology were, however, challenged by a significant number of respondents. Various individual and collective actors stated their concerns regarding the possible undesired consequences that GMOs may have on the natural and social environments. There appears to be a homophony among participants who opted for a more cautious approach to the cultivation and commercialization of GMOs on at least five major themes:

• There is *no* scientific consensus on the safety of GMOs;
• While GMOs *may* display some benefits, these are seriously overstated by GM advocates;
• The precautionary principle should be sustained as an informing principle of risk assessment;
• There are serious concerns that GM technology is—and will be— appropriated by large corporations that create an oligopolistic environment;
• The focus on GM technology, in essence, undermines the consideration of alternative agricultural innovations and techniques.

As is obvious, some of the themes developed by GM skeptics are the binary opposites of the ones backed by GM advocates, while others stretch the discussion to different subject areas and, in this way, highlight the richness of the GM controversy. Participants who expressed their concern about the potential negative impact that GMOs may have on human health, the natural environment and social order include among others political bodies such as the governments of Northern Ireland, Scotland and Wales; research centers such as Social, Technological and Environmental Pathways to Sustainability (STEPS); non-governmental environmental organizations such as Greenpeace and GM Freeze; associations, organizations and charities of the civil society such as the Soil Association, the Family Farmers' Association, "Which?" (the largest European consumer organization with almost 800,000 members) and the Alliance for Natural Health International and individual scientists with specialties in the field, such as Professor Brian Wynne, (Emeritus Professor of Science Studies at Lancaster University and vice chair of the Food Standards Agency's (FSA) Steering Group on GM; Professor Andy Stirling, professor of Science and Technology Policy at Sussex University and codirector of STEPS center; Professor Paul Nightingale, professor of Strategy at the University of Sussex and deputy director of the Science Policy Research Unit (SPRU).

The hearings were completed on January 7, 2015, and almost a month later the STC published the report *Advanced Genetic Techniques for Crop Improvement: Regulation, Risk and Precaution*, which assessed the viewpoints of individuals and associations that had submitted written and oral evidence.

Some of the conclusions reached by the STC appear to be in tandem agreement with the arguments of GM advocates and include the following:

- There are "major flaws" in the EU's regulatory system that threaten to prevent GM products from reaching the market (House of Commons Science and Technology Committee 2015a, 3);
- Allegations of scientific uncertainty regarding long-term effects of GM crops cultivation are *not supported* by the available scientific evidence and "should not be used as a pretense for value-based objections" (House of Commons Science and Technology Committee 2015a, 17–18);
- GM technology is not a cure-all for global agricultural problems, but it has a role to play—along with various technological, social, economic and political approaches—in meeting the challenge of sustainable and secure global food production (House of Commons Science and Technology Committee 2015a, 20–21);
- "The Government's approach to agricultural research is balanced and does not focus excessively on genetic techniques" (House of Commons Science and Technology Committee 2015a, 24);
- The STC has not been convinced that the application of intellectual property rights to GM crops has hindered other innovation trajectories, and it has not been presented with enough evidence to support claims that patents hinder independent research. Since, however, this is a complex matter, further consideration is warranted (House of Commons Science and Technology Committee 2015a, 27).

The government welcomed the committee's report as "a valuable contribution to the debate," and which offered "a considered and challenging perspective on how advanced genetic plant breeding techniques should be addressed in policy and regulatory terms" (House of Commons Science and Technology Committee 2015b, 1). The main concerns expressed during the inquiry regarding the safety and negative ramifications GMOs may have for the natural and social environments—ramifications that were doubted in the STC report—were eventually dismissed in the governmental response. It was clearly communicated that "The Government's general objective is to reduce the burden of regulation and to ensure that any necessary controls are pragmatic and proportionate. To support innovation products made from safe technologies must have a clear [...] access to the market" (House of Commons Science and Technology Committee 2015b, 1).

A Brief History of the GM Debate in the UK

Despite the polyphony of arguments articulated from the opening statements of the STC inquiry until the final decision of the British government, there is a single thread that can connect all viewpoints and demonstrate that, despite the lack of consensus, there is still a common denominator among engaged parties. The need to reframe the GM debate and encourage public involvement was expressed virtually unanimously and was, in turn, endorsed by governmental representatives as an initiative that needs to be implemented imminently. This, however, inevitably raises the questions: What kind of dialogue does one desire and what sort of public involvement can be pursued?

For example, the very inquiry being discussed came immediately under fire from STEPS, which challenged the neutrality of the whole endeavor by openly questioning "the prejudicial formulation of the Inquiry and the partial and leading questions listed" which, if not reformulated in a less-partisan way, would "risk undermining the scientific and democratic legitimacy of the Select Committee's work in this Inquiry and more generally" (STEPS Centre 2014). While the criticisms of STEPS were directed to the STC, the John Innes Centre offered a firm reply to the former's accusations. STEPS's suggestion that the inquiry was "single-mindedly promoting GM" was dubbed as a "caricature" (John Innes Centre 2014, para. 8) and it was further argued that, while STEPS should have welcomed this inquiry, they failed to offer material contributions to the questions raised by the inquiry regarding the harm arising from GM crops (John Innes Centre 2014, para. 1). Professor Wynne also expressed his skepticism toward the narrow way in which the specific inquiry, but also previous initiatives, had framed the problematic. "The framing basically is always GM or nothing. That is not what we have got, and it is not what scientific research should be basically bolted down to either" (House of Commons Science and Technology Committee 2014, Q.61).

The National Consensus Conference

The STC inquiry was the latest in a series of public-engagement initiatives on agricultural genetic modification that have taken place in the UK over the past two decades (House of Commons Science and Technology Committee 2015a, 61). On November 2, 1994, the first lay panel on plant biotechnology, the "National Consensus Conference on Plant Biotechnology," was held. It was funded by the Biotechnology and Biological Sciences Research Council (BBSRC), despite the council's initial reluctance, and was organized by the Science Museum "whose staff implicitly diagnosed the problem as public misunderstanding or anxiety" (Levidow and Carr 2010, 118). The panel was regarded as "an experiment in democracy" and was heralded as a response to "the perceived inadequacies of representative democracy" (National Centre for Biotechnology Education 2006). The panel consisted of 16 lay volunteers who set the agenda for the conference and chose the expert witnesses called to attend and to reply to the panel's questions. The report produced by the panel—which was deemed "exceptionally measured and balanced" by John Durant the assistant director of the Science Museum—recognized that "biotechnology could change the world" and that plant biotechnology in particular "has a role to play in helping to provide the world with quality food, and with non-food products from sustainable sources" (National Centre for Biotechnology Education 2006, Question 1). Nonetheless, the panel recognized that "[t]he impact of plant biotechnology on the environment is extremely difficult to predict," as they received the conflicting opinions and information from what they discerned as "the environmental and scientific lobbies" (National Centre for Biotechnology Education 2006, Question 3). As a result, despite giving to "the field of plant biotechnology its qualified support," the panel made specific suggestions aimed at mitigating the potential risks of GM technology such as: the strengthening of regulations regarding the release of GMOs into the environment; the establishment of effective international controls over the commercialization of GM plants; and the provision to consumers of clear and comprehensible information about new biotechnological products (National Centre for Biotechnology Education 2006, Preface). The overall impartiality and openness of this "experiment in democracy" appeared to be unconvincing on numerous occasions. Before the conference started, two members of the steering committee attempted unsuccessfully to exclude from expert status representatives of what were considered to be extreme anti-biotech groups and, consequently, from the list prepared by the organizers. As the conference progressed it became obvious that the chairman tended to give pro-biotech speakers the entitlement to appear knowledgeable on diverse aspects, while NGO activists were put on the defensive to demonstrate their expertise; finally, after the hearings had come to an end, concerns about who would legitimately steer biotech innovation, who would appropriate agbiotech and who would be accountable for environmental monitoring were marginalized and appeared in the report as issues of safety and patent control (Levidow and Carr 2010, 118, 9; Purdue 1995, 1996).

The GM Nation? *Public debate*

In May 2002, the government announced that it was sponsoring an unprecedented national debate on the genetic modification of plants: the *GM Nation?* public debate. The government's intention was to use this exercise to gauge public understanding of issues related to agbiotech and, most importantly, to "take public opinion into account as far as possible" in the decision-making for future GM policy (Levidow and Carr 2010, 126). The public debate unfolded in parallel with two other strands of *GM Nation?*: a scientific and an economic. The former was carried out by the Chief Scientist's GM Science Review Panel, which reviewed the literature relevant to risk assessment, and the latter was assigned to the Prime Minister's Strategy Unit, which conducted a cost–benefit analysis of GM technology. A few months before the public aspects of the debate were to commence, however, the EU and, as a result, the UK—bound by European law—had restarted the approval process of GM products. The timing of this development was unfortunate, and the chair of the *GM Nation?* Steering Board wrote to the secretary of state seeking reassurances about the government's good will and sincere intentions (Horlick-Jones et al. 2007, 5, 6). Despite the government's efforts, the debate unfolded amidst a feeling of distrust as many participants were sympathetic to the belief that the government's commitment to take the results of the debate seriously was a hollow one (Horlick-Jones et al. 2007, 168).

The findings of the public debate strand of *GM Nation?* showed that public skepticism, which had been discerned in the 1994 National Consensus Conference, had now acquired more backbone. It was found that participants had "strong anxieties about risks from GM, particularly towards the environment and human health" (Department of Trade and Industry 2003, 6) and, as a result, preferred that "clear and trusted answers to unresolved questions about health and the environment" were given before commercialization of GMOs was given the green light (Department of Trade and Industry 2003, 7). The incentive driving large agbiotech corporations to promote GM technology was also questioned, as it was recorded that "[p]eople believe that these companies are motivated overwhelmingly by profit rather than meeting society's needs, and [...] they have the power to make their interests prevail over the wider public interest. [...]" (Department of Trade and Industry 2003, 7). The findings of the Science Review Panel did nothing to allay the public's concerns, but most likely verified their validity. The possibility that GM plant -reeding techniques may have unexpected outcomes was clearly stated in the Chief Scientist's report and so was the likelihood of toxic or nutritionally deleterious effects resulting from cultivation and consumption of GMOs, which cannot be completely ruled out (GM Science Review Panel 2003). Regarding the cost–benefit analysis conducted by the Strategy Unit, the results were no less discouraging for the potential economic benefits of GM technology. In particular, the study concluded that the public's attitudes would play a decisive role in determining the commercial success of GM foods, and there was scant evidence that the current generation of GM crops—that is, the first generation of GMOs, the only one commercially available at the time of writing—would bring any economic benefit to the UK (Strategy Unit 2003, 16).

The ill-fated FSA public dialogue

The third, and thus far final, public engagement project was launched in 2009 when the Labour government commissioned the Food Standards Agency (FSA) to launch an $800,000 public dialogue over GM food in order to gauge the public mood on agribiotechnology. An independent steering group was appointed to shape the project, make key decisions about how the dialogue would be designed, approve all materials used in it and oversee the project's evaluation (Food Standards Agency 2011). The endeavor, however, was abruptly terminated when, in June 2010, two steering group members resigned. One member was Dr. Helen Wallace, director of GeneWatch UK, the first to resign in protest at the FSA's links with the agrichemical industry, which had been lobbying for the broad commercialization of GM foods. "It has now become clear to me that the process that the FSA has in mind is nothing more than a PR exercise on behalf of the GM industry," she said. The second member to resign, Professor Brian Wynne (vice chair of the FSA's steering group, professor of Science Studies at Lancaster University, and associate director of the ESRC Centre for Economic and Social Aspects of Genomics) also accused the FSA of having adopted a "dogmatically entrenched" pro-GM position (Vidal 2010). In his letter of resignation, Professor Wynne ends by noting:

> [T]he evaluation which the currently favored contractor candidate is proposing to do, is most unlikely to take the learning from this exercise beyond existing inadequate understandings of public dialogue on such science-intensive but by no means solely, indeed not primarily, scientific issues. They are public issues involving science, not 'scientific issues', and this evaluator team showed little if any sign of understanding that, nor the issues and problems I have only indicated above[…] I joined this process because I felt there was a chance to help explore these deeper problems with relevant policy institutional actors; but I feel now, in the light of experience and reflection, that this was unduly optimistic. (Wynne 2010a)

The resignations stirred considerable media coverage and, in September 2010, Science Minister David Willetts announced the government's decision to halt the project in its current form and take "this valuable opportunity to step back and review past dialogues on GM and other areas of science to ensure [they] understand how best to engage the public over such issues" (qtd. in *Sciencewise* 2010).

The "Second Push" of the Agbiotech Industry

While it is difficult to determine the extent to which the UK government actually took into consideration the concerns voiced by the public, it is teleologically evident that, by and large, the government concurred with the favorable scientific advice it received on GMOs. The influential *Reaping the Benefits* report, published in 2009 by the Royal Society, stressed the pivotal role that GMOs have to play in meeting the global challenges of food security, increasing world population, climate change and growing scarcity of water and land (The Royal Society 2009). The report was endorsed by the Council for Science and Technology (CST), which alluded to its findings in a letter addressed to the prime minister following the latter's inquiry on the risks and benefits of GM technologies (Council for Science and Technology 2014). The pro-GM advice of the CST to the prime minister was based not only on the recommendations of the Royal Society but also on the *GM Science Update* report commissioned by the CST for a group of independent scientists chaired by Professor Baulcombe. The significant overlap between the two reports should, however, come as no surprise, as four of the five authors of *GM Science Update* were also in the working group involved in producing the *Reaping the Benefits* study. These include Professor Sir David Baulcombe (University of Cambridge), Professor Jim Dunwell (University of Reading), Professor Jonathan Jones (John Innes Centre) and Professor John Pickett (Rothamsted Research). In essence, the second report supplemented the findings of the initial study, as almost half a decade had gone by since the publication of the first, and new technoscientific advancements had become a reality. In the section "New scientific development over the last five years," the scientists' enthusiasm for GM technology is pervasive. "The potential for new GM crop varieties is likely to increase greatly [...] we expect more complex novel traits in GM crops including the production of novel compounds for biofuels and industrial use. Such GM crops could be key components of an expanding bio-economy[...] To realize the full potential of GM crops we need to improve the European regulatory and approval process and to strengthen the R&D pipeline" (Baulcombe et al. 2014, 2–4).

In the spirit of realizing the full potential of GM crops, a number of initiatives took place by which political bodies and the agbiotech industry collaborated in tandem so as to help unlock, as it has often been argued, the economic potential of the UK's world-leading agricultural research. On June 26, 2012, Science Minister David Willetts, Lord Taylor, George Freeman MP, and Roger Williams MP participated in a roundtable meeting with an agenda based on the ABC's *Growing for Growth* report. Academics from UK universities and research institutes, such as Rothamsted Research and the John Innes Centre, were also present, along with representatives of the National Farmers Union (NFU). Among others, the summary of the meeting, publicized by ABC, revealed plans to increase investment in biotech—particularly in R&D, and in technology that will help farmers—and focus more on plants and biotech on the syllabus at all educational levels. As obstacles that confine the growth of agbiotech, the *Going for Growth* roundtable feedback pointed towards: (a) the regulatory barriers and political divisions at national and EU level; (b) the limited availability of funding for investment and education; (c) the structure of agriculture, which is made up of a vast number of small farms looking at different sectors, thereby rendering commercialization difficult. Finally, regarding the actions that have to be taken in order to change the existing problematic situation, it was suggested that the government should, for its part, make improvements in the regulatory framework, offer incentives for investment, have a clear position to take to Europe; research institutes and academia should forge a mutual understanding of what is at stake and offer a clear response to anti-GM groups; the private and public sector should forge better overall cooperation on agbiotech R&D (Agricultural Biotech Council 2012).

In October of the same year the UK Department for Business, Innovation & Skills (BIS) published a call for evidence on "Shaping a UK Strategy for Agri-Tech." The aim of the Agri-Tech Strategy was to deliver "a clear, shared long-term vision for unlocking the full economic potential of the UK's world-leading research [and] focus on technology across the agricultural sector from the research laboratory through the food supply chain[…]" (Department for Business, Innovation & Skills 2012, 5–6). The responses received to this call for evidence helped shape the development of the *UK Agricultural Technologies Strategy*, launched on July 22, 2013 (BIS 2012). The new strategy ratified the partnership between government and industry in "improving the EU regulatory framework and unlocking new opportunities in trade and investment" (BIS 2013, 3). The program's overarching aim was to bring the UK "at the forefront of the global race to sustainable intensification" by allocating $250 million government investment in the Agri-Tech Strategy, which would be "led by industry, working in partnership with the public and third sectors" (BIS 2013, 3–4). Such "partnerships between industry, the science base and Government will come into their own" as multiple strategies—including innovative engineering and novel approaches to crop and livestock genetic improvement—are implemented (BIS 2013, 13).

The Rothamsted GM wheat field trials

However, reports on the benefits of GM technology did not simply rely on future projections. The work carried out by Rothamsted Research in general, and the 2012 GM wheat field trials in particular, very often spearheaded the pro-GM argumentation and posited as a fine example of ground-breaking research being carried out on British soil.[4] GM skeptics, however, framed the Rothamsted field trials as part of the "second push" orchestrated by the biotech industry to get GM into the UK (Take the Flour Back 2012b) almost ten years after the British public kicked GM technology out the door (Channel 4 2012).

Although there was no commercial growing of GMOs in the UK in 2012, two experimental field trials of GM potatoes (one carried out by the University of Leeds and the other by the Sainsbury Laboratory) and one trial on GM wheat (performed by Rothamsted Research) were granted authorization that year (DEFRA 2010b). Of the three field trials, the Rothamsted Research experiment received the most significant attention by the media, with headlines ranging from an unequivocally "pro-GM" stance (Connor 2012) to clearly more-skeptical and cautious opinions (Clover 2012). Despite a respectable number of GM skeptics who firmly expressed their views publicly, it appeared that for the very first time in the UK, media coverage seemed to be positive towards the possibilities that agbiotech allegedly can offer. In fact, of the 22 editorials published between May 21st and June 1st, 2012 in publications that have been traditionally doubtful of agbiotech, 18 were clearly favorable to GM while only four were against (Moses 2012).

The aim of the experiment was to assess whether wheat genetically engineered by Rothamsted Research would be able to repel aphids—serious pests of wheat and other arable crops that transmit viruses and reduce yields—by producing the $E\beta f$ pheromone. If the GM trials produced as successful results as the ones seen in the laboratory, the benefits to farmers and the environment would be manifold as insecticide spraying would be reduced and farming would be made more sustainable. The study was funded entirely by the BBSRC, on behalf of the UK public, and the total costs for the research project, specifically, amounted to $1.1 million. What skyrocketed the cost, however, was the amount invested in fencing and additional security measures. In order to protect the trial site from intruders and wild animals, the BBSRC invested an additional $700,000 and, in fear of vandalism and criminal damage, the BBSRC offered an additional $2.8 million (Rothamsted Research 2015).

Quite obviously, what attracted the attention of the media was not only the scientific significance of the field trials but, most importantly, the public stir that was caused. The research project was faced with opposition from a number of groups, organizations and individual GM skeptics. On March 1, 2012, the organization GM Freeze launched the campaign GM Wheat? No Thanks! in order to oppose the Rothamsted experiment. The campaign was supported by NGOs and groups such as GeneWatch UK, Greenpeace, the Gaia Foundation, EcoNexus, Real Bread Campaign, Bakers Food and Allied Workers Union and many more (GM Freeze 2012d). While the campaign orchestrated by GM Freeze was basically about raising awareness of the issues surrounding GM wheat and, eventually, urging individuals and groups alike to "send a strong message to the Government and regulators" that GM wheat (trials) should be banned (GM Freeze 2012c), another campaign was much more radical in scope. Take the Flour Back, a grassroots network of individuals that had no membership and no authorized representatives, saw the field trials as a "real, serious and imminent contamination threat to the local environment and the UK wheat industry" (Take the Flour Back 2012c). Although GM Freeze and Take the Flour Back shared a lot of the criticisms raised against GMOs, there appeared to be a divergence as to how the "threat" posed by GMOs should be dealt with. Instead of opting for institutionalized means of protest, Take the Flour Back warned Rothamsted that if the research center and the government did not "remove the GM plants themselves" by May 27, 2012, the grassroots network would have to follow the "only avenue open" to them, which was to "decontaminate" the area (Take the Flour Back 2012). Although the network and Rothamsted engaged in various short debates in *The Guardian*, on Channel 4, and on BBC Radio, the central debate proposed by Rothamsted in a neutral venue (Friends Meeting House) and with a neutral chair (George Monbiot) never took place, as Take the Flour Back declined the invitation, arguing that they had not been notified early enough (Rothamsted Research 2012e; *Guardian* 2012).

Amidst fears of vandalism, scientists from Rothamsted Research launched the "Don't Destroy Research" appeal and petition that constituted a plea from scientists and individuals from all over the world to make activists reconsider their alleged plans to destroy years of scientific labor. More than 6,000 individuals signed the petition (Sense about Science, 2012) and some 44,000 people became aware of Rothamsted's appeal by watching the video uploaded by the institute (Rothamsted Research 2012h). On the morning of May 20, a man (who was eventually charged with criminal damage) broke into the Rothamsted Research Center. "The intruder caused significant, random property damage, but failed to disrupt the experiment in this attack" (Rincon 2012). Later that day, a cyber attack took down Rothamsted's website for a few hours (BBC 2012). The planned direct action against the GM wheat experiment did not take place on May 27th. Although some two hundred Take the Flour Back demonstrators gathered near the field plots, no attempt of "decontamination" took place. The protesters walked towards the site but were stopped by police lines. Although two arrests were recorded, the police said there were "no disturbances in the area" (BBC 2012).

Despite those attempts to "decontaminate" the site, the field trials progressed successfully and, in the summer of 2015, the results were announced. To the disappointment of the research team involved in this five-year project, the GM wheat had failed to repel aphids in the field. The difference in aphid infestation between the GM wheat and the conventional wheat used as control was statistically insignificant. Huw Jones, senior biologist at Rothamsted, with oversight for the genetic changes in the plants said: "We had hoped that this technique would offer a way to reduce the use of insecticides in pest control in arable farming. As so often happens, this experiment shows that the real-world environment is much more complicated than the laboratory" (qtd. in Rothamsted Research 2015).

The significance of the field trials

While the experiment proved of little practical importance, as plans for commercialization have been abandoned, the symbolic and theoretical implications of the field trials are of considerable salience. Several obvious realizations support this suggestion. First, the field trials acted as a lightning rod for the wider GM controversy as the sequence of themes dominating the broader debate reappeared within the context of the Rothamsted study. Issues that informed the discussion of the STC inquiry—the three endeavors for public dialogue and the numerous reports submitted to the government, such as the long-term effects of GMOs on human health and the environment, the problem of patenting and corporate control, the relevance of the precautionary principle and the effectiveness of UK and EU regulations—all reappeared in the discussion of the wheat trials. Second, the experiment under discussion can offer a concrete understanding of how social dynamics are interrelated and operate on national and international levels since the UK is bound by EU legislation and the GM trials unfolded within such as legislative framework. Third, from a theoretical/methodological standpoint, the study of the wheat experiment can prove to be an ideal entry point to the rich discussion of GMOs, as the research can start from a modest scope (i.e., a specific experiment carried out within a clearly defined time/space location), where the particularities of the trials are examined and then broaden the analytical horizon by tracing issues that are anchored across different time/space locations.

The Need for a Holistic Framework

If the Rothamsted experiment is accepted as a socially significant conjuncture in its own right and as a convenient passageway to the broader GM debate, then the question that immediately arises is how this research project should be theoretically and methodologically approached. Before turning to this more challenging problem, certain preliminary remarks may help facilitate the task at hand. From the discussion on GMOs developed so far in this chapter, it is rather unproblematic to observe the following:

- The GM debate is not an aggregate of free-floating events, but a conjuncture embedded in larger historical and sociocultural processes;
- At the core of the conjuncture lies a scientific issue that needs to be clearly addressed. This, however, should not be realized by undermining the importance of economic, political/legislative, cultural and social forces in the shaping of the controversy;
- Actors from these different institutional spheres vie for the control/appropriation of the technology by using institutional and/or non-institutional means. Some such methods include the practice of patenting, the creation and amendment of legislative frameworks, the implementation and revision of methods of risk assessment and risk management, and acts of "decontamination," vandalism and boycotting;
- Ideology not only appears to play an important role in the way that GM technology is construed, but there is an undeniable ideological rift between GM advocates and skeptics;
- Specific actors are involved in the conjuncture with their own subjective worlds (i.e., dispositions, motives, capabilities, interpretation of the external environment, psychological makeup and so on) which affect and are affected by the emergent circumstances;
- Involved actors come from different institutional spheres (i.e., political, economic, cultural, social and scientific) and are located on different strata of social hierarchies. In this way, they contribute unevenly to the production and reproduction of social life by being able to exert influence and mobilize resources in varying degrees.

By acknowledging that, at the very least, the above broad observations need to be accommodated in the process of social analysis, certain corresponding theoretical and methodological guidelines should now be broadly sketched. The field trials and the wider controversy need to be examined via a holistic framework which:

- Offers a balanced and non-deterministic exegesis of how forces from different institutional fields intermesh across various social locations;
- Neither compartmentalizes nor conflates the interactive relation between scientific knowledge and the social order;
- Critically assesses how various actors try to appropriate technology;
- Takes into account the ideological elements underlying the controversy and examines what these reveal about the broader dynamics of the debate;

- Situates actors within specific position–practice relations and examines the asymmetrical distribution of power;
- Can go beyond the description of events and perform hermeneutic and structural analyses.

The social aspect of agricultural biotechnology is highlighted in a growing number of books and edited volumes.[5] These works offer invaluable insights into specific aspects of agribiotechnology, such as the ethical and social dimensions of GMOs, the political economy involved, the formation of networks of influence, the legislative frames in different continents of the world, the eruption of collective protest and more. However, this is done without anchoring the empirical evidence within a holistic and systematic theoretical frame that would offer a broader understanding of trajectories taking shape beyond the microcosm of the carefully gathered data of each case study. This book aims at complementing the field's nascent literature by demonstrating that a critical understanding of agbiotech is feasible with the use of key sociological concepts, coherent ontological classifications and appropriate methodological bracketing.

The overarching aims of the book, therefore, are threefold, with each strand intertwined with the others. First, to propose a way of filling the analytical gap found in the current literature by offering an original theoretical framework. Second, to examine the 2012 GM wheat field trials carried out by Rothamsted Research, which have been associated with the "second push" of agbiotech firms to bring GMOs to the UK, by using the suggested framework. Third, to examine the key elements underscoring the GM debate and suggest a reframing of the controversy, which moves beyond the simplistic conceptualization of it being a case of science versus politics.

The structure of the book

In order to critically assess the wheat trials and appraise the social dynamics of the wider debate from a holistic perspective, the theoretical and methodological foundations of the framework need to be laid first. For this to happen in a clear and coherent manner, in Chapter 2 the discussion focuses on the major approaches in science, technology and society relations, while in Chapter 3 the major strands of science and technology studies (STS) are assessed. A table with a cumulative overview of all the major ontological and methodological points suggested by the various approaches is provided in order to make it easier for the reader to reflect on theories; a task which can often be quite demanding and complex.

In Chapter 4, the contributions of Benton, Mouzelis and Stones to contemporary sociology are presented. Each theorist's work, relevant to the task at hand, is discussed not only on its own merit but also in relation to the problems highlighted in the two preceding chapters. In Chapter 5, the suggested holistic framework, a critical synthesis informed by the work of these three sociological thinkers, is offered as a promising way of examining the field of agribiotechnology from an externalist and an internalist perspective. To this end, methodological brackets are highlighted as essential tools that allow the investigation of research questions across varying levels of ontological abstraction and scale in a substantive and meaningful way.

Chapters 6 and 7 signify the return to the empirical investigation of the field trials and revisit the issues touched upon in this introductory chapter to a greater depth. By demonstrating the heuristic value of the framework, the two chapters shed light on the key aspects of the Rothamsted experiment, at the same time exploring the broader institutional arrangements, key ideological constructs and the social order.

Chapter 8 starts by evaluating the typical conceptualization of the GM controversy as a binary case of science versus politics. In assessing the validity of this argument, key themes of the conjuncture are discussed: whether a scientific consensus on the safety and benefits of GMOs exists; the promises of first and second generation GMOs and whether they have been fulfilled; the case of the Golden Rice, and more. The second part of the chapter suggests a radical reconceptualization of the debate, which takes into account the environmental character of GM technology and the multitude of stakeholders involved. By stressing the merits of public engagement, a particular type of sustained dialogue that can complement formal decision-making frameworks is tentatively fleshed out.

Chapter 2

RETHINKING SCIENCE, TECHNOLOGY AND SOCIETY RELATIONS: DEFINITIONS, BOUNDARIES AND UNDERLYING THEORETICAL PROBLEMS

In order to better understand how GM technology can affect our social lives, and how social actors—like ourselves—can, in turn, shape the trajectory of technoscientific innovation, we must first take a logical step backward and inform our viewpoint from the main approaches to science, technology and society relations. More specifically, this chapter discusses the concepts of technological determinism and social constructivism, Thomas Misa's "meso-level approach," Thomas Hughes's "technological momentum" and Sheila Jasanoff's idiom of "co-production." As the purpose of this chapter is to offer the reader a clear understanding of key concepts in science, technology and society relations—which they can refer to as the discussion on GMOs unfolds in the ensuing chapters—I use the main text for the development of the major arguments and use footnotes for the analysis of finer points. While not an exhaustive set of approaches to the investigation of the interplay between science, technology and society, these five viewpoints constitute the most prominent and distinct avenues for the study of the problematic under discussion. By the end of this chapter, it is hoped the reader will have become familiar with each approach's main tenets and will have also acquired a critical understanding of the strengths and weaknesses of each.

Technological Artifacts, Scientific Knowledge and the Social Order

What is technological determinism?

While the term technological determinism is used quite extensively in science and technology studies (STS), it is a construct that one comes across not only in theoretical works, but it can be easily discerned in the discourse of everyday life in many popular phrases or beliefs: "the automobile created suburbia," "the robots put the riveters out of work," "the Pill produced a sexual revolution," "the personal computer changed the world of politics and business"[1] and so on. In all such statements, complex events are construed as inevitable results of one single technological innovation. In effect, advancing technology is seen as a steadily growing, well-nigh irresistible, force that determines the course of events (Smith and Marx 1994, xi, xii).

In academic works, facets of technological determinism are often discerned in the works of Lynn White, Jacques Ellul and Karl Marx. For example, White's account that feudal society was the result of the invention of the stirrup is one of the most oft-cited illustrations of technological determinism. "Few inventions have been so simple as the stirrup, but few have had so catalytic an influence on history […] The Man on Horseback, as we have known him during the past millennium, was made possible by the stirrup" (White 1964, 38).

French philosopher Jacques Ellul has also been extensively engaged with the idea of technological determinism and, in a passage that has striking resemblance with "economic determinism of the last instance," he argues that "I am by no means saying that technology has always, and in all societies, been the determining factor [but] that in our Western world (and we can generalize for the past twenty years), technology is the determining factor" (1980, 67).

In some very influential papers, such as those of Robert Heilbroner (1967, 1994), Hughes (1994) and Langdon Winner (1977), Karl Marx is also portrayed as an advocate of technological determinism.[2] It is argued that Marx as a technological determinist is revealed in some of his oft-quoted passages, such as: "The hand-mill gives you society with the feudal lord; the steam-mill, society with the industrial capitalist" (Marx 2001, 119); or "modern industry, resulting from the railway system, will dissolve the hereditary division of labour, upon which rest the Indian castes, those decisive impediments to Indian progress and Indian power" (qtd. in Bottomore 1991, 75).

Despite the prevalence of the term technological determinism in popular culture and STS works, there is little consensus regarding its meaning and its underlying causes (Bimber 1990, 333–34). In order to avoid unnecessary confusion, a good starting point for clarifying the concept under discussion can be found in Cohen's seemingly truant,[3] but still apposite, remark that "technological determinism is, presumably, two things: it is technological, and it is determinist" (Cohen 2000, 147, fn1). Elaborating on that simple distinction, Bruce Bimber argues that an account that can legitimately be classified as technological determinism "should hold that history is determined by laws or physical and biological conditions rather than human will. This makes it deterministic.

And it should rely upon features of technology to explain those determining conditions or laws. This makes it technological" (1990, 340).

This twofold distinction appears to be crucial and is sustained in one form or another in the works of Ted Benton (1994, 34, 36), Allan Dafoe (2015, 1052), Donald MacKenzie (1984, 474, 476), MacKenzie and Judy Wajcman (1999, 3–5), Misa (1994, 120) and Merritt Smith and Leo Marx[4] (1994, 2). All these theorists seem to agree that the concept of technological determinism is couched on two fundamental beliefs. The first is the idea that technological progress follows an autonomous trajectory that is insulated from economic, political, cultural and social pressures (i.e., what Cohen and Bimber classify under *technological*). In this way technology is considered to develop and move along axes created from its very own inherent logical structure (Svensson 1979, 305) or that "technology appears autonomously to beget technology in general" (Beniger 1986, 9–10). The second constitutive element of technological determinism is the idea that technological innovations determine social change[5] (i.e., what Cohen and Bimber relate to *determinism*).

The fundamental flaw in technological determinism

The main problem with the concept under discussion lies in the fact that the technology–society bond is regarded as a one-way equation (technology → society) while the opposite relation (society → technology) is, by definition, simply impossible to theorize. In theoretical terms, this is a typical problem of *reductionism*. Reductionism is observed in instances in which the complexities of social life and the process of social change are attributed to a single, unifying, prime factor that instigates a cause: in our case, technology. These prime movers can be "the interests of the middle class," "risk society," "globalization," "anxiety" and so on. Reductionism is very likely to appear when a theorist ascribes to certain analytical items/categories (such as the ones mentioned) a de facto causal primacy. While it is certainly true that, on some occasions, the causes of a phenomenon can indeed be either one or two, problems arise when this is done in an a priori fashion *before* empirical research is carried out to justify the chief causes. In other words, reductionism is apparent when "explanations of social phenomena [give] undue emphasis to some causal forces at the expense of other significant forces that are correspondingly denied their due causal autonomy and efficacy" (Stones 2005, 40).

A reductionist mode of theorizing, however, is not the only problem with technological determinism. Nicos Mouzelis's fruitful discussion on the problems of *system* essentialism and *actor* essentialism can help shed more light on the notion under discussion. For this Greek sociologist, the former refers to instances in which actors either disappear from the theoretical framework—or are portrayed as mere "placeholders"—and structures are endowed with anthropomorphic qualities. Accounts of system essentialism can be traced in works that explain social order and social change by reference to general and impersonal "systemic needs," "structural functions" and, in our case, "technological progress." Actor essentialism, on the other hand, is observed when some theorists reach the other extreme and—by ignoring institutional realities—portray actors as rational decision-makers[6] with pre-constituted interests, motives and identities (Mouzelis 2008, 237–42). Actor essentialism can be discerned in the theses of "the historical mission of the proletariat," "the ruling class ideology" and so forth.

The major problem with technological determinism, therefore, is that it is couched in a reductionist understanding of the social world—an understanding that does not, and cannot, accommodate the multiplicity of factors shaping social life (economic, political, social and cultural)—and, at the same time, commits the fallacy of system essentialism by depriving actors of the significance of their social actions and their "power to do otherwise."

Beyond technological determinism: The case of social constructivism

While it is true that a considerable number of theorists refute technological determinism for being a reductionist mode of thinking, it is equally true that the reasons they do so are quite diverse. At the other extreme from technological determinism lies *social constructivism*, which questions the omnipotence of the "machine" and stresses the importance of human actors. A significant volume of work in the field has, directly or indirectly, challenged the two sub-theses of technological determinism; (a) that technology has its own internal logic, and (b) that its effects irrevocably shape social structures. The supposed naturalness and inevitability of technological trajectories are convincingly doubted in studies that range from the British TSR2 aircraft (Law and Callon 1992; Law 2002) to fluorescent lighting (Bijker 1992) and the construction of aircraft (Schatzberg 1999). What these studies have in common is what Wiebe Bijker (1995), David Bloor (1973), Trevor Pinch and Bijker (1984) and Sally Wyatt (2008) call the "Principle of Symmetry." This means that "success" or "working" is the *result*, not the cause of a machine becoming a successful or working artifact. In other words, technological advances do not follow a unique, linear and evolutionary trajectory, but their success is subject to the pressures/interests of social groups, political decisions, the economic context and various cultural elements. What these works, which broadly speaking fall into the school of social constructivism,[7] also suggest is that among a series of products that are formed inside the technoscientific field, the ones that finally emerge victorious are not necessarily the best in terms of efficiency or effectiveness.[8]

Although these works fall into the paradigm of social constructivism and totally reject technological determinism, their accounts stem from two different perspectives, namely: *humanism* and *post-humanism*. The humanist stance places human agents at the center of discussion and construes technologies as objects, as passive recipients of the agents' actions. Post-humanism, on the other hand, "decenters the subject" and places humans and nonhumans on a par by claiming that differences between human beings and living organisms or machines are ontologically and methodologically nonexistent. The most notable advocates of humanism include Harry Collins, Steven Yearley, Bijker and Bimber among others while some of the most prominent endorsers of post-humanism are Bruno Latour, Michael Callon, John Law, and to a certain extent, Andrew Pickering (Breslau 2000).

The differences between humanism and post-humanism are not limited to the ontological or theoretical levels but have significant methodological implications. Humanists examine the social environment and follow humans and interest groups in order to offer an understanding of how a technological artifact was produced by depicting, at the same time, technology and machines as dead raw material. "For humanists, it is all about the subject. Humans are center stage" (Matthewman 2011, 18). Post-humanists reject this methodological approach by deconstructing the boundaries between human and nonhuman and suggest that a more appropriate approach is to construe both as "actants." "There are no humans in the world. Or rather, humans are fabricated—in language, through discursive formations, in their various liaisons with technological and natural actors, across networks that are heterogeneously comprised of humans and nonhumans who are themselves so comprised" (Michael 2000, 1).

Some central problems in social constructivism

Since the writings of these theorists are extensively discussed in the next chapter, at this point I only raise two fundamental objections against humanism and post-humanism. First, although their starting points are very different, both perspectives imply, or in some cases clearly state, an infinity of trajectories that technological artifacts may take. This is so because humanism construes technologies as "passive or mechanical effects of human will" (Breslau 2000, 300), and the latter ascribes an "infinite pliability" to technological artifacts (Latour 1997, 4).

Regarding this "infinity" of trajectories, I am much more inclined to agree with Heilbroner's thesis (1967) on the predictability of technology, which argues that technologies may move through a *finite* number of avenues.[9] It is not only Heilbroner's analysis that is more convincing, but it is the writing of some humanists and post-humanists that also inadvertently point toward that direction. To be more precise: in their seminal study of bicycles Pinch and Bijker (1984) argue against the linear model of technological determinism and instead claim that a "multidirectional" model of technological trajectories is closer to reality. "Such a multidirectional view is essential to any social constructivist account of technology. Of course, with historical hindsight, it is possible to collapse the multidirectional model onto a simpler linear model[,] but this misses the thrust of our argument that the 'successful' stages in the development are not the only possible ones" (Pinch and Bijker 1984, 411). It appears, therefore, that there is a contradiction between the theoretical and the empirical parts of most social constructivist studies. On the one hand, technological artifacts are seen as "dead material" shaped according to human will, and they go undertheorized for large parts of the analyses but, on the other hand, they are presented in such a way that only a *huge but finite* number of trajectories may be followed. The initial theoretical assertion that technologies are "passive effects of human will" is not sustained by examples of technological developments as products of rampant human imagination. Instead, only a specific number of possibilities is presented, and the reason this finite spectrum is available is not discussed at all.

The same line of criticism as to why it is that *certain* artifacts appear can be raised with respect to the studies of Law and Callon (1992). In this case too, the "infinite pliability" of the theoretical level is nowhere to be found on the empirical level, where the TSR.2 aircraft could undergo a *finite* number of different modifications and at the same time the British government was presented with a limited number of aircrafts to choose from.

The second objection to humanism and post-humanism is that their way of theorizing human–nonhuman relations leads either to social determinism (social forces "design,"[10] which trajectories technology will follow) or to a transcendence of the problematic. As a result, what is offered in the place of technological determinism is either a very weak, malleable form of technological product, or a "hybrid" form of human–nonhuman interactions. Therefore, the relation of technology–society either takes the form society → technology or goes untheorized by arguing that the two cannot be distinguished from each other. By giving primacy to social forces in an a priori way the same unconvincing answers that technological determinism offers are given, with the difference that the terms technology–society are now swapped. The conflation of the two terms is equally unsatisfactory as, for one thing, gaining insight about technological artifacts via a hybrid terminology puts the researcher at the risk of resorting to anthropomorphism (Benton and Craib 2001, 71).

The rest of this section assesses three distinct alternatives to technological determinism and social constructivism offered by Misa, Hughes and Jasanoff. The first suggests a meso-level network analysis, the second theorizes the concept of "technological momentum" and the third refers to the idiom of "co-production." All three approaches try to avoid the two extremes by arguing that technology both shapes and is shaped by society.

Misa's meso-level approach

As a distinguished historian, Misa has extensively written on the history of technology and technological determinism (1988, 1992a, 1992b, 1994). In his influential article "Retrieving Sociotechnical Change from Technological Determinism," Misa attempts to explain the reason some theorists espouse technological determinism while others discern social forces behind technological artifacts. For him, the adoption of either of the two competing perspectives is related to the level of analysis a theorist opts for. In this way, a macro-level study is "more prone to technological determinism," while a micro-level study is "apt to find more contingent and multiple societal forces at work in the historical process" (Misa 1994, 115). But what does Misa conceptualize with the micro–macro distinction?

> In distinguishing between the macro and the micro, it is essential to emphasize that the issue is not merely the *size* of the unit of analysis. Besides taking a larger unit of analysis, macro studies tend to *abstract* from individual cases, to impute *rationality* on actors' behalf or posit *functionality* for their actions, and to be *order-driven*. Accounts focusing on these 'order-bestowing principles' lead toward technological, economic, or ecological determinism. Conversely, accounts focusing on historical contingency and variety of experience lead away from all determinisms. Besides taking a smaller unit of analysis, such micro studies tend to focus solely on case studies, to refute rationality or confute functionality, and to be disorder-respecting. Generally, macro studies make it easy for historical actors to appear rational, purposeful, and as key agents of change, whereas micro studies make it difficult or impossible for historical actors to have the same attributes. (Misa 1994, 119, stress in original)

Two remarks can be made about Misa's argument that technological determinism is contingent on the level of analysis that the researcher opts for. First, contrary to Misa's claim, micro-level studies are not incompatible with the rationality of actors and the functionality of their actions. In fact, the whole paradigm of rational-choice theory is about micro-interactions and the de facto rationality of actors. In rational-choice theory, actors are all about careful calculation, reflexivity, monitoring and anticipation of their actions' consequences (Goldthorpe 2000). Second, while Misa is right in pointing toward a tendency of macro-level studies to demonstrate serious affinity to deterministic accounts, as it was already developed in the discussion of technological determinism, this actually has very much to do with the imbalanced theorizing between or among actors and systems and not with the level of analysis. Despite the strong tendency that Misa correctly discerns, there is certainly nothing inevitable about it; such methodological shortcomings can be avoided so long as they are properly identified.

In order to avoid the extremes of technological determinism and social constructivism, Misa suggests an alternative meso-level approach. He believes that in order to "formulate a new and more insightful analysis of technology and social change, historians should direct attention to what can be called the 'meso' level—the region conceptually intermediate between the macro and the micro[; ...] this means analyzing the institutions intermediate between the firm and the market or between the individual and the state" (Misa 1994, 139).

This methodological leap to the meso level is offered as a solution to reductionism or system/actor essentialism. What appears to be Misa's argument is that a meso-level approach will allow the researcher to focus on more situated processes and, hence, on more contingencies. Therefore, the argument goes, the focus on social interactions of the meso level can accommodate a closer investigation of both the actors' true motives and technology's influencing role.

While, in principle, Misa's proposition appears to offer a practical way out of the two extremes of determinism and constructivism, it does not seem to constitute either a necessary or a sufficient methodological approach. This is so because it is still not very clear how the conceptual region between the firm and the market or the individual and the state helps the researcher grasp the dynamics between technology and social change. In effect, there are absolutely no conceptual guarantees that such a meso analysis will deter the theorist from ascribing to cartels or investment banking houses rationality and purposefulness of action that, according to Misa's syllogism, lead to determinism. So long as there are no explicit methodological *and* ontological guidelines as to how technology–society relations should be approached, the errors Misa so rightly stresses can very likely reappear at a meso-level trajectory, too.

Although Misa makes a very lucid and useful distinction by assessing technological determinism's two sub-theses, the alternative approach he offers would most likely benefit from further elaboration. The meso-level approach he suggests may help the researcher avoid the stances of technological determinism and extreme social constructivism only as long as certain methodological guidelines and ontological typifications are clearly spelled out, so that the problems of system and actor essentialism are convincingly tackled.

Hughes's "technological momentum"

Hughes's concept of "technological momentum" constitutes another alternative to technological determinism and social constructivism (Hughes 1994, 102). Hughes finds unsatisfactory these two extremes, which argue that "technical forces determine social and cultural changes," and that "social and cultural changes determine technical change" respectively, and in their place offers the idea of technological momentum (ibid.). Technological momentum suggests that social development shapes and is shaped by technology, and the extent to which this happens is time-dependent (1994, 102). But what is Hughes's concept about, and how does it help the researcher eschew technological determinism and social constructivism?

Hughes's argument is that technological systems both shape and are shaped by society under certain circumstances. More specifically:

> A technological system can be both a cause and an effect; it can shape or be shaped by society. As they grow larger and more complex, systems tend to be more shaping of society and less shaped by it. Therefore, the momentum of technological systems is a concept that can be located somewhere between the poles of technical determinism and social constructivism. The social constructivists have a key to understanding the behavior of young systems; technical determinists come into their own with the mature ones. Technological momentum, however, provides a more flexible mode of interpretation and one that is in accord with the history of large systems [...] It suggests that shaping is easiest before the system has acquired political, economic, and value components. It also follows that a system with great technological momentum can be made to change direction if a variety of its components are subjected to the forces of change. (Hughes 1994, 112)

The technological momentum thesis suggests, therefore, that technological systems are more easily shaped by social forces when they are small and less complex in terms of bureaucracy, technical or physical infrastructures and, in turn, when they grow larger and become more complex, these systems are more likely to affect social structures (Hughes 1994, 113).

While Misa focused more on *methodological* bracketing in order to avoid the two poles of technological determinism and social constructivism, Hughes appears to be placing greater emphasis on *ontological* issues to achieve the same goal. First, he underlines how important a role the size and complexity of a system play and, second, he attempts to shed more light on various aspects of the technology–society interplay by recognizing an internal division of labor within the technological systems and discerning technical and social components. There is no doubt that Hughes is correct in stressing the need for an ontology that accommodates both social and technical elements and places them in a society that is not flat, but consists of systems of varying size, complexity and influence. One cannot, however, help but feel that Hughes's terminology, and subsequent conclusions, leave quite a few crucial questions unanswered.

It seems that some fundamental problems with the "technological momentum" thesis are couched on its very building blocks (i.e., the definitions of social, technical and technological systems).

> Privileging of the technical in a technological system is justified in part by the prominent roles played by engineers, scientists, workers, and technical-minded managers in solving the problems arising during the creation and early history of a system. As a system matures, a bureaucracy of managers and white-collar employees usually plays an increasingly prominent role in maintaining and expanding the system, so that *it then becomes more social and less technical.* (1994, 105–6, stress added)

My objection is not targeted to the arbitrariness of the more social and less technical classification of systems (e.g., how would one convincingly classify Microsoft, General Motors and Apple as "more social and less technical" systems?), but to the ways this change of balance within a system is materialized in the first place. How are the "prominent roles played by engineers, scientists" and so on eventually relegated with the advent of "white-collar employees and managers?" Who gave them prominence in the first place? Was it an automated change or were there actors behind this shift in power and responsibility? In other words, would it not be much more insightful to analyze an organization in terms of its internal division of labor—the relation among departments, allocation of funds for marketing, R&D, and so on—instead of resorting to the "more or less" "technical" or "social" terms? Would not a closer analysis of a system reveal better how technologies are designed, produced and reproduced within a system by focusing on a more *situated ontology* that deals with specific scientists, particular apparatuses and so forth?

Apart from these theoretical objections, there is empirical evidence that does not quite verify the technological momentum thesis. Hughes suggests that as time goes by and systems become larger and more complex, they are less likely to yield to environmental pressures; although as he concludes "technological momentum [...] is not irresistible" (1994, 113). The case studies Hughes uses do demonstrate his point, but there are, however, striking examples of "systems" with huge market shares and significant financial leverage that were forced to adapt to the environment regardless of their size and complexity. I have in mind cases such as Microsoft's anti-trust trials (BBC 2003), Apple's import ban (Decker 2013), and General Motors' filing for bankruptcy protection (Isidore 2009). Such cases, just to name a few, clearly demonstrate that it is a fallacy to construe that a complex system has primacy over its environment and, therefore, ignores the complexities of the social world.

To conclude, there is no doubt that Hughes is right in pointing out that the two poles of technological determinism and social constructivism can be evaded so long as an ontological distinction between social and technical is sustained, and the way that the two interact is brought to the surface. Nonetheless, the fact that he does not elaborate enough on his own terminology leads him, in the end, to retreat to the two extremes he wanted to avoid. It seems that a more substantive ontological analysis is necessary if one is to avoid convincingly giving a priori primacy to actors or technologies.

The idiom of "co-production"

The concept of co-production is of great relevance to the study of technoscientific innovation, and agbiotech in particular, as it is applied across diverse domains of research in STS and is gaining wide recognition as an alternative approach to deterministic accounts (Irwin 2008, 589; Jasanoff 2006b, 2). Co-production is not a fully fledged theory claiming consistency and predictive power, but rather constitutes "an idiom—a way of interpreting and accounting for complex phenomena so as to avoid the strategic deletions and omissions of most other approaches in the social sciences" (Jasanoff 2006b, 3). The strategic deletions and omissions Jasanoff is referring to are the extremes of technoscientific and social determinism (Jasanoff 2006a, 20) already discussed in this chapter, and their tendencies to ascribe primacy to technoscientific artifacts or the social order, respectively. As a response to these challenges, the idiom of co-production stresses the "*constant intertwining of the cognitive, the material, the social and the normative*" (Jasanoff 2006b, 6, stress added). In this way, it sensitizes the researcher to investigate "the simultaneous formation of social and natural order in knowledge societies" (Jasanoff 2008, 772). Though not offering substantive research guidelines, co-production suggests that the natural and social orders intermesh in four distinct pathways: in the formation of *identities* as roles with power and meaning; in the production and reproduction of *institutions* as a web of social and normative understandings; in the development of *discourses* as markers of boundaries between the promising and the fearsome aspects of technoscientific innovation; and in the creation of technoscientific *representations* as intelligible schemata articulated in diverse communities of practice (Jasanoff 2006a, 38–41).

Being aligned with the interpretive and post-structuralist turn in the social sciences (Jasanoff 2006a, 38), co-production shares fundamental ontological claims with ANT and particularly with the principles of agnosticism and "generalized symmetry," as we clearly see in the following chapter. In order to avoid repeating points that will be more fully developed in the discussion of ANT as a fully fledged research model with its ontological claims and methodological guidelines, only a few epigrammatic remarks will now be made. First, it is true that the shaping of identities, institutions, discourses and representations is contingent not only on normative and social factors but also on material and cognitive parameters. It is also true that the birth, production and reproduction of technoscientific artifacts is an outcome of the ways these four elements intermesh. Having said this, the way these four components of social and natural reality intertwine cannot be coherently examined on an ontological level of abstraction, which tends to conflate these components at the expense of underlining their distinct characteristics. Second, and following from the previous point, rigorous analysis can be achieved so long as the level of ontological abstraction becomes more specific, and particular cognitive schemata, material properties, institutional arrangements and normative frameworks are assessed. By taking this step toward a more situated ontology, it becomes obvious that social and natural entities cannot be placed on a par, but need to be systematically distinguished according to the function they have under clearly defined time/space locations. One can distinguish, for instance, between the material arrangements that constitute an actor's *conditions of action* (e.g., the lab kit that a researcher has at their disposal or the tools a farmer can use), the material properties of a particular artifact and the amount of *conjuncturally specific knowledge* a scientist has about those, the social and cultural order that prevails during a given period of time and the way this has been internalized or drawn upon by an agent-in-situ and so forth. In this sense the claim that "[n]atural and social orders [...] are produced at one and the same time" (Jasanoff 2005, 19)—which closely resembles Giddens's duality of structure thesis, examined in Chapter 4—is rather misleading. It would be more accurate to suggest, it seems to me, that the natural and social orders are *present* at the same time as conditions of action, as resources the agent can draw on or as integral parts of the agent, but which may acquire *different roles* as an action unfolds. Some elements may become means, other ends, and other outcomes of an action that may eventually become crystallized as new environments of action. I turn to these points in Chapter 4.

To summarize, these points raised here are not aimed at challenging the excellent points that the idiom of co-production makes. On the contrary, it is argued that they need to be safeguarded against conflationist tendencies, doing so via a framework that accommodates them within a situated and discrete ontology. We can now turn to reiterate the most important points raised in this chapter.

Technoscientific progress and social change: Backward and forward

- Technological determinism is observed in studies that explicitly or implicitly argue:
  a. Technology is autonomous and follows its own linear trajectory;
  b. Technological innovation dictates social change.
- Technological determinism is not necessarily related to the micro–macro distinction, but is primarily a problem of reductionism or *system* essentialism.
- While social constructivists (humanists or post-humanists) are correct in rejecting technological determinism, the alternatives they offer are no less problematic:
  a. Humanists, in the end, espouse social determinism, which is a corollary of reductionism or *actor* essentialism;
  b. Post-humanists transcend the dichotomy by conflating humans and nonhumans into hybrid "actants," thus making it impossible to theorize technology–society relations.
- Technology–society relations have to be theorized through a framework that
  a. takes into account how human agents and larger social forces (economic, political and cultural) shape technological artifacts;

b. does not construe technologies as "dead" and "infinitely pliable" material, but recognizes that technological artifacts may follow a huge but *finite* number of trajectories;

c. applies appropriate methodological brackets that help the researcher distinguish between various levels of social reality;

d. rests on ontological classifications that help capture the rich and complex ontology of social life without conflating or transcending fundamental distinctions.

Chapter 3

SCIENCE AND TECHNOLOGY STUDIES: A CRITICAL OVERVIEW OF THE FIELD

In Chapter 1, the main approaches to the relations between science, technology and society were discussed. These viewpoints sketched the broad ways in which the interplay between society and technoscientific progress can be conceptualized. In this chapter, the discussion focuses on the major theoretical strands in science and technology studies (STS)—a sociological sub-field that was born in the late 1930s and remains as one of the discipline's "marginal specialties" (Shapin 1995, 289). Two things should be clarified about the nature of this chapter. First, the review offered is certainly not exhaustive of the field, but rather highlights the most germane approaches in STS. Second, it is targeted to a critical discussion—to the fullest extent possible—of each approach's theoretical, ontological and methodological premises. The reason for this lies not only in the aim of this book, that is to build gradually an informed holistic framework in STS, but also rests on the fact that all too often an overview of the literature tends to yield toward the philosophical rather than the sociological foundations of each approach; often giving the impression that this sub-field of sociology remains uninformed of certain developments in contemporary sociological theory and ignorant of key writings of the past.

There is a general consensus that the first substantial work in the sociology of science and technology appeared in 1938 (Knorr-Cetina 1991). Merton's classic monograph, *Science, Technology, and Society in Seventeenth-Century England*, heralded a new era in the field of sociology as it pried open the gates of the fields of science and technology and invited sociologists to appraise a domain that up to then was accessible only to natural scientists. Since then, the sociological study of science and technology has evolved to such an extent that many subbranches have emerged. The study of scientific knowledge (SSK) and, more recently, science and technology studies, and Actor-Network Theory (ANT) form some of the broad main avenues in this sociological enterprise. STS has become an intellectually influential interdisciplinary field which, besides academics, attracts the interest of activists, scientists, decision-makers (from the political and economic spheres), doctors and so forth (Hackett, Amsterdamska and Lynch 2008, 1). As is the case with other fields of sociology, in this area too there is little consensus among scientists regarding the way its dynamics can be grasped.

Back to Merton

Although Merton is considered to be the founder of the sociology of science (Knorr-Cetina 1991), with the exception of his article on the ethos of scientists (Merton 1942), his work is either ignored in contemporary STS or simply portrayed as one of the first unsuccessful attempts in the study of science. The aim of this section is to demonstrate that Merton's contribution is still significant and, if this is duly appreciated, some major problems in influential STS approaches can be avoided. Toward this goal, I provide as concrete an analysis as possible of the various links that Merton drew between science and the other institutional spheres. I start by referring to Merton's PhD thesis and then trace how the recurrent theme of science's relative autonomy was developed in a number of books and articles that the American sociologist published throughout his prolific life.

The place of science in society

In his *Science, Technology and Society in Seventeenth-Century England*, Merton claims that one of the reasons the field of science evolved during that particular period in England was because of the spirit of Protestantism. This happened because of the normative patterns that the latter evangelized and the former endorsed. To be sure, Merton never discerned any *causal* or *deterministic* relationship between the two institutional spheres by claiming that the Puritan ethic *caused* science or that *only* Puritans were scientists. Rather, he pointed out that, on the one hand, science found a very fertile ground thanks to the values of Puritanism and, on the other hand, Puritans found in their scientific pursuit an activity that embodied Puritan teachings (Merton 1973, 224).

Some years later Merton elaborated on his initial thesis by expanding his argument well beyond the case of Puritanism and seventeenth-century England. For him, the interdependence of various institutional spheres, in general, is unquestionable.

> Socially patterned interests, motivations, and behavior established in one institutional sphere—say, that of religion or economy—are interdependent with the socially patterned interests, motivations, and behavior

obtaining in other institutional spheres—say, that of science [...] The same individuals have multiple social statuses and roles: scientific and religious and economic and political. This fundamental linkage in social structure in itself makes some interplay between otherwise distinct institutional spheres even when they are segregated into seemingly autonomous departments in life [...] Separate institutional spheres are only partially autonomous, not completely so. It is only after a typically prolonged development that social institutions, including the institutions of science, acquire a significant degree of autonomy. (Merton 1973, 175)

Although in this case we have a broader claim about the place of science in society than in his initial monograph, the nature of the linkage is basically the same. Merton is still arguing that *one way* the different institutional spheres are related to one another is via the norms and values that actors acquire/internalize from the religious sphere, for example, and that accompany them in social interactions that unfold in different social settings.

Another type of linkage between science and society, however, is revealed in his 1938 article, *Science and the Social Order*. In this case, the relationship is not related to *internalized* norms that actors carry with them across various social settings, but is contingent on *external* influence stemming from different social spheres. As an example, Merton assessed the situation in Nazi Germany, where politically imposed Aryan criteria dictated who was allowed to enter universities and laboratories and, therefore, the educational and scientific spheres in general. The rise of Nazism saw the political arena questioning the independence of another institutional field and, eventually, depriving it of its relative autonomy. "Scientists, as well as all others, [were] called upon to relinquish adherence to all institutional norms that, in the opinion of political authorities, conflict with those of the State" (Merton 1938, 325). Therefore, the ethos of science[1] (intellectual honesty, integrity, organized skepticism, disinterestedness, impersonality) was sacrificed by racial and political criteria.

In his foreword to Barber's *Science and the Social Order*, published in 1952, Merton tried to formulate a *third* type of linkage between the scientific field and the other spheres; unfortunately, this linkage was more of an insinuation rather than a developed argument. In that short piece of writing, the American sociologist attempted to depart from his normative functionalist framework and tried to point toward avenues of influence between science and society that existed well beyond the presence of norms. In that work, the analysis shifted to the way the social and cultural/religious context of a society impinges upon the scientific field. For Merton, science is an organized social activity and "the measure of [this] support and the type of scientific work for which it is given differ in different social structures, and the directions of scientific advance may be appreciably affected by all this" (1952, 216). This *external* pressure does not question scientists' motives and does not imply that science is "the mere appendage of political, economic, and the other social institutions" (Merton 1952, 216). While Merton does not elaborate further on how social games played at other institutional spheres affect the scientific field, he is pointing toward a competing school of thought at the time—thought that could eventually shed more light on the issue. Merton rather bitterly admits that the relation between science and society is not taken into account by his contemporary theorists and argues that although Marxism has interesting things to say about this relation, the realization of such a fruitful discussion was highly unlikely at the time (Merton 1952, 217). While it would be rather audacious to surmise what Merton had in mind, he is clearly advocating that a Marxist account—which obviously stresses the influence of the economic sphere in ways that are much more versatile than the levying of norms—could shed more light on the ways that the "direction of scientific advance" is shaped, so long as it does not merely resort to the base/superstructure dichotomy.

Finally, in his article *The Unanticipated Consequences of Social Action*, published in 1936, one can trace a *fourth* link between science and society. This relation does not examine how society influences science, but how the scientific field may have an impact on society. In examining the factors that lead to unanticipated/undesired consequences of action, Merton mentions, among others, the idea of "imperious immediacy of interest." This refers to "instances where the actor's paramount concern with the foreseen immediate consequences excludes the consideration of further other consequences of the same act" (Merton 1936, 178, 179). For Merton skepticism or even hostility toward science stems from the fact that there is a discrepancy between the methodological views (field's norms) and the social results of science. The belief that scientists should be focused "exclusively on the scientific significance of their work with no concern for the practical uses to which it may be put or for its social repercussions generally"—because doing so will result in the possibility of bias and error—is the reason that lies behind the fact that certain technological advances (such as the atomic bomb) have had devastating effects on humanity (Merton 1938, 329, 330). In this way, he clearly expresses his concerns about the groups of people that will control these technologies. Whether intentionally or unintentionally, the notion of "imperious immediacy of interest" and the ways in which this can create conflict opens up the possibility of a Marxist account (which he very much welcomed in the 1952 article mentioned above) that can delve deeper into the notion of "appropriation" and the ways in which actors from different institutional spheres vie to get hold of technological artefacts.

Merton: Backward and forward

To conclude, it seems that with careful reading, Merton's work on the sociology of science has much more to offer than it is usually given credit for. The American sociologist was the first to conceptualize science as a field that could be assessed from a sociological standpoint. He did so by acknowledging the place of science in society as a sphere that enjoys relative autonomy; in other words, as a sphere characterized by a distinct set of norms and values, but which is, nonetheless, affected in a variety of ways by the other major institutional spheres (economy, polity, culture/religion). Merton also developed his understanding of the ways science and society interact. For him, the science–society linkages can materialize the following ways:

- By actors who have internalized norms from one field (i.e., polity, economy, culture) and carry them during interactions that unfold within the settings of the scientific field and vice versa. In the scientific field, these include intellectual honesty, integrity, organized skepticism, disinterestedness and impersonality;
- In instances in which an institutional sphere (e.g., polity/military or religion) has primacy over the others (as in authoritarian or theocratic states respectively), and it imposes on the scientific field its own norms and values, threatening the field's relative autonomy;
- By the technological artifacts that the scientific field produces, and can affect, the other institutional spheres in both anticipated and unanticipated ways.

It seems to me that all of the above points that Merton makes are not only valid but should be seriously taken into account. Placing science in society as an institutional sphere that both affects and is affected by the other fields is a first broad theoretical gesture that helps the researcher start to realize that *social* games are played in the scientific field too. Nonetheless, it equally seems to me that in order to unravel the dynamics of these interactions, the foundations laid by Merton need further elaboration. The main areas that would most likely benefit from further analysis are the following:

- The portrayal of actors as merely carriers of norms; there are definitely more parameters in an actor's social life that should be taken into account, such as their dispositions, access to resources, how they are placed in relation to other actors, how they accommodate the potential conflicts of norms they may have internalized and more;

- The more or less ad hoc presence of technological artifacts—the way technologies are designed, developed and controlled needs a further breakdown. While Merton did not engage with the task of theorizing how technologies are shaped, produced and reproduced, he clearly pointed toward the Marxist concept of "appropriation" within the relations of production. Merton's implication could lead toward a very interesting and revealing trajectory.

Science as a Golem: The Sociology of Scientific Knowledge

As seen in the previous section, Mertonian sociology of science focused on the normative and institutional arrangements that enable science to exist as a field of relative autonomy. For Merton, the content of scientific knowledge remained a closed book. On the other hand, sociologists who comprise the, largely British, specialty called the "Sociology of Scientific Knowledge" (SSK) have been "concerned precisely with what comes to count as scientific knowledge and how it comes so to count" (Collins 1983, 267). This radical shift in focus was signaled in 1976 by David Bloor's book *Knowledge and Social Imagery*. From the opening lines, Bloor made his intentions very clear.

"Can the sociology of knowledge investigate and explain the very content and nature of scientific knowledge? Many sociologists believe that it cannot […] I shall argue that this is a betrayal of their disciplinary standpoint" (1).

The origins of the "Strong Programme" in SSK

In this work were laid the foundations for what Bloor coined the "Strong Programme" in the sociology of scientific knowledge. Bloor's aim was to examine whatever counts as knowledge in a particular culture and that culture's social characteristics (Yearley 2005, 21). In order for a social scientist to assess scientific knowledge, Bloor argued, four tenets have to be adhered to: *causality*, *impartiality*, *symmetry* and *reflexivity*. First, SSK has to be *causal* in that it should be "concerned with the conditions which bring about belief or states of knowledge." Second, SSK has to be *impartial* on truth and falsity, success or failure, rationality or irrationality; that is, both sides of these dichotomies will have to be explained. Third, SSK explanations have to be *symmetrical* in the sense that the same types of cause would have to explain success and failure. Finally, SSK needs to be *reflexive* to the extent that the patterns of explanation it uses are applicable to sociology itself (Bloor 1976, 4, 5).

The overall thrust of the Strong Programme's four tenets was that knowledge should be studied with the same tools and with the same explanatory end in mind. "The thesis of methodological symmetry does not presuppose that the categories of truth/falsity, reasonable/unreasonable are bogus. The claim is rather that it is methodologically undesirable to make use of the differential assessment of belief in developing a naturalistic account of belief transition" (Newton-Smith 1981, 251). Nonetheless, many scholars criticized Bloor's thesis on the grounds that it, supposedly, considered *all* knowledge to be the same.

> Much of the controversy which followed the book's publication was conducted at a hysterical level. Philosophers, natural scientists, anthropologists and psychologists joined the dozens of sociologists who, at conferences and in reviews, unleashed their "ultimate refutations" of Bloor's work. Bloor even found himself sharing lecture platforms with parapsychologists and other cognitive deviants, invited to philosophers' conferences as an epistemological freak-show. (Yearley 2005, 22)

Despite such incidents of intentional or unintentional misinterpretation, the Strong Programme "inspired a large number of sophisticated and insightful empirical studies of science" (Benton and Craib 2001, 62). The new variants of SSK that have sprouted since then include studies that examine scientific knowledge and/or technological artifacts from the viewpoint of discourse analysis, ethnomethodology, reflexivity, social interests and gender, among others (Yearley 2005).

The Empirical Programme of Relativism: Legacy and main tenets

In order to try to capture *some* of the richness that can be found in the field of SSK, in this sub-section I focus on Harry Collins's Empirical Programme of Relativism (EPOR), since Collins "has been first and foremost among those who have taken the content of natural scientific work seriously, and by doing so, have turned the sociology of science around and refocused it on 'SSK'" (Knorr-Cetina 1998, 491). It is rather obvious that EPOR occupies a very special place in SSK, and it seems to me that the reason is twofold. First, it is a very representative approach of the field because EPOR is a direct descendant, so to speak, of the Strong Programme; Collins's approach, formulated in 1981, retained two of the four Strong Programme's tenets (i.e., symmetry and impartiality). Second, EPOR is still a very dynamic approach nowadays, used to demonstrate the social aspect of scientific, technological and medical advancements/controversies (Collins and Pinch 1998, 2002, 2005).

In order to understand what the acronym EPOR is all about, we first have to make sure that its two basic integral parts—that is, *"empirical"* and *"relativism"*—are clearly defined. Collins's principal interest is in natural scientific knowledge, not in mathematics or logic and, for him, in a scientific dispute or controversy what is precisely up for grabs are the facts of the matter (Yearley 2005, 28). However, for Collins, "the natural world has a small or nonexistent role in the construction of scientific knowledge" (1981, 3). If the natural world plays either a small or a trivial part in the way scientific progress is made, it follows that EPOR's "endorsement of relativism means that it must seek to explain the content of scientific knowledge as far as possible in social terms" (Collins 1983, 272). This has been exactly the focus of EPOR scholars in the past decades. Since the 1980s researchers following the Empirical Programme have offered to the field inspiring sociological insight as to how scientific controversies regarding gravitational radiation, the chemical transfer of memory, the theory of relativity and the origins of life (Collins and Pinch 1998), just to mention a few, have all ended because a series of "negotiations" among scientific actors took place (Pickering 1992, 1). EPOR, therefore, suggests that the relativistic aspect of knowledge should be revealed through studies of a highly practical/empirical orientation.

For Collins, there are three stages in EPOR:

- Demonstrating the interpretative flexibility or inevitable openness of scientific results;
- Describing the social mechanisms that *limit* interpretative flexibility and thus close debates over results;
- Investigating how the consensual interpretation of laboratory work and results is possible within constraints coming from *outside* the community of scientists (Collins 1981a, 4,7; Yearley 2005, 29).

These three parts are EPOR's overarching elements, which argue that social and cultural factors are the primary forces in closing scientific controversies. It has to be noted, however, that the presence of all three stages does not constitute an obligatory passage point in an EPOR study. The third phase, in particular, has been the one to cause the greatest problems for social analysis (Collins 1981a, 7; 1983, 276) and has been quietly dropped in the later works of Collins and Pinch (1998; 2005).

The contribution of SSK to the field of sociology cannot be stressed enough. To repeat, following the foundational work of Merton on the birth and autonomy of the scientific field, theorists who followed Bloor and his ground-breaking Strong Programme offered the discipline an impressive spectrum of studies ranging from the discovery of the magnetic monopole and the chemical transfer of memory to the "missing" solar neutrinos and the theory of relativity. The first training of several of these researchers was in natural science, and this made it easier for them to become familiar with the technical details of the area of science under investigation. By conducting in-depth interviews and engaging in participant observation in the scientific activity under analysis, researchers were able to investigate the *content* of scientific knowledge as opposed to its institutions alone. All these examples, which fall under the approach of EPOR, demonstrated the undeniable *social* facet that underlies the interpretation of experiments and the closure of scientific controversies.

It seems, however, that the social aspect of scientific knowledge has been stressed to such an extreme that certain features, functions and characteristics of the scientific, technological and natural world have been unjustifiably downplayed. At times, EPOR appears to be a rather lopsided construct, favoring social construction and attributing quasi-Promethean qualities to human action, while at the same time nature, science and technology serve as an adorning backcloth to unfolding social interactions. It appears to be the case that EPOR's *empiricism*, *relativism* and *symmetry* cause issues that affect almost every facet of the approach. I discuss all of these problems in turn.

The problems of empiricism and relativism in EPOR

As we have already seen, EPOR rests on two pillars; empiricism and relativism. To repeat, empiricism in EPOR argues that we should base our theories and beliefs about scientific knowledge and controversies on empirically grounded case studies. Empirical research on specific issues, usually on the frontiers of science, help grasp the exact ways that breakthroughs are achieved and debates are closed. Relativism, on the other hand, suggests that our beliefs about the natural world are to a great extent, if not entirely, independent of that world. To echo Collins's relativistic stance once again: "the natural world in no way constrains what is believed to be" (1981b, 54). What EPOR is arguing, therefore, is that empirical studies in science, technology and nature confirm the thesis of relativism; that is, that science, technology and nature have little or no bearing on our beliefs about them. While this is a methodological claim and not an ontological one, I think that Larry Laudan (1982) is right in pointing out that it is very difficult for empiricism and relativism to sit well together.

> The search for evidence in any field of empirical inquiry is the search for statements which reflect some features of the world. The very notion of evidence is relational. Evidence is always evidence *of something* or other. Unless we believe there is some linkage between a statement and a certain state of affairs in the world then we refuse to regard the statement as evidential. But if the thesis of strong relativism were correct, it would be pointless—even self-contradictory—to cite evidence for that thesis,

since strong relativism denies that there is an evidential relation between our assertions and the world. Indeed, to cite empirical evidence for any claim is to concede that strong relativism is misguided, precisely because strong relativism denies the relevance of empirical evidence. (131–32, stress in original)

In the conclusion of his seminal work on gravitational waves, Collins remarks:

I hope that the detailed empirical work found in this paper, and in other papers in the same tradition, will bear out the claims made for the relativistic approach and encourage its adoption as a methodological prescription, even by those to whom it is epistemologically distasteful. (1981b, 54)

It is not very clear how the impressively thorough empirical work that Collins has to demonstrate in the field of scientific knowledge is actually compatible with the relativistic approach he encourages other researchers to adopt. If facts of the natural world are so methodologically irrelevant and peripheral as Collins advocates, then on what grounds did the British sociologist choose to interview and follow the work of these *particular* scientists and the findings of *those* laboratories? Collins's choice "makes sense only on the assumption that there are certain facts of the matter, that these facts are at least partially accessible to us, and that these have a role to play in delimiting one's theories about the subject matter" (Laudan 1982, 132).

In the carefully crafted empirical studies that Collins presents, he is very cautious not to censor actors or ascribe to them motives, intentions knowledge and so on without the appropriate empirical evidence. When Collins is not in the position to present data that support his claims, his remarks very often take the following form: Scientist A *may* have done so and so, actor B *could* have thought that […], it would make sense to *imagine* that the debate continued in that way […], and so on. Collins, in other words, does not commit the methodological fallacy of attributing to actors characteristics necessary to verify a narration decided a priori by the researcher; a point raised by Misa and Rob Stones in the previous chapter. Nonetheless, while Collins is very discrete in observing social interactions unfolding, he does so within a methodological frame that has already been compromised. Collins's a priori commitment to relativism appears to be a source of concern. To be clear, the issue here is not that Collins and EPOR put an emphasis on the social world by bracketing the natural world; this kind of methodological bracketing would not pose a problem at all. The issue is that the worth of scientific, technological, and natural elements is purposively *downgraded*. It is tough not to regard the position that the natural world has little, or no, significance to scientific knowledge as a clear facet of sociological reductionism. Collins, quite naturally, does not discern any form of reductionism in this viewpoint and challenges his critics: "If writers wish to soothe themselves and others by endorsing an 'objective' world, they should make clear exactly where they think it must intrude in accounts of scientific practice—preferably with examples" (Collins and Cox 1976, 438). What Collins is actually asking from his opponents is to provide examples of science reflecting a purely "objective" external reality that is devoid of social interpretations. "Of course, Collins is sending us on a fool's errand. Presumably, the only counter-evidence that he would consider damaging would be examples of where the 'facts' of the natural world unmediated by social processes determined the outcome of scientific debates over 'truth'" (Gieryn 1982, 288). Collins's challenge implies a nature/society dichotomy in which the two are, more or less, mutually exclusive elements. It seems that Collins's insinuation is rather misleading. "What makes science unique, in part, are institutionalized procedures which define the intersection of natural and social worlds. The appropriate question is not *if* the natural world intrudes in scientific constructions of knowledge, but *how* it does so in science in a different way than in religion or the arts or even common sense" (Gieryn 1982, 289, stress in original).

The fact that Collins construes the relation between scientific knowledge and the natural order as nonexistent or, at best, ambivalent is reflected in his equally ambivalent opinion on science and everyday life.

> As we have seen in the bombing scandals, contested forensic evidence is like contested scientific evidence everywhere; it is like the science described in this book. It is contestable [...] Doubts about evidence can always be raised. But it does not follow from this that forensic evidence should carry no weight. In judging the merits of forensic evidence we have to apply the normal rules which would apply if we were judging any argument between experts. For instance, some experts will have more credibility than others and some will have no credibility at all. (Collins and Pinch 1998, 145)

It is evident that Collins is not willing to yield to any relativistic aphorisms of scientific knowledge *in toto*, but it is not very clear either what is keeping him from doing so. Why does forensic evidence carry some weight? Is it because some experts are more credible than others? On what merit do some experts become more credible than others? Is it because their evidence carries more weight? It appears that Collins is begging the question. There is certainly nothing in EPOR's relativism that would warrant Collins's claim. In fact, Collins is giving some credibility to scientific knowledge, not because of EPOR, but *despite* the approach he has established over the years.

The problem of symmetry in EPOR

The absence of a clear-cut stance on the credibility of scientific findings appears to stem from the principle of *symmetry* that EPOR abides by. To repeat, Collins's empirical program has retained two of the four principles suggested by Bloor: namely, impartiality and symmetry. Impartiality suggests that in SSK all beliefs or knowledge require explanation, and symmetry argues that these beliefs or knowledge should be treated equally and assessed using the same analytical tools. While the criterion of impartiality does not pose any concerns as it helps expand the field of study to beliefs that are not necessarily true, the criterion of symmetry seems to be problematic. This is so because symmetry fails to reflect the way that scientific reasoning is developed. For Newton-Smith (1981), Collins does not acknowledge the simple fact that beliefs or scientific knowledge can be judged according to what he calls the "dictates of reason."

> When what someone would offer as his reason for believing p does indeed provide reason for believing that p, I will say that he is following the *dictates of reason*. If someone is following the dictates of reason, then showing that this is so […] explains his belief. If he is not following the dictates of reason we shall, *ex hypothesi*, have to give a different type of explanation for his believing what he does. Failures to follow the dictates of reason can be divided into those that are rationalizations and those that are not. The latter would include cases of carelessness, lack of intelligence, lack of interest, and cases in which the person in question is acting on a hunch and cannot provide any further reason. (254, stress in original)

What Newton-Smith argues, rightly I think, is that while all beliefs and knowledge require explanation, the explanations given are not symmetrical. For example, Hesse (1986) argues that while Collins is right in stressing the importance of experiments, he is telling us only half the story. "Collins speaks as though replicability of individual experiments were the only recognized criterion for scientific validity [but] replicability has never been the working scientist's sole criterion of acceptable science" (718, 719). Other standards for the evaluation of experimental results include methods of data analysis, coherence between facts and theory, development of worldviews and more. "That is, arguments directly related to the experiment are always mixed with theoretical, aesthetic and pragmatic criteria" (Godin and Gingras 2002, 142). Other critics such as Allan Franklin (1994) and Sylvia Culp (1995) also stress the importance of the scientific community and the argumentation used within it to end controversies. All of these scholars stress the fact that scientific debates are closed on *epistemological* grounds, which for some reason Collins neglects to mention.

In light of the discussion developed so far in this chapter, the last sentence of the above quotation echoes like a clear reference to the fact that Collins's analysis would make perfect sense if it were resituated within the scientific field that Merton has so extensively discussed! Could it be that a scientist's credibility is not a social criterion, but a product of *prior* contribution to the field through publications, inventions, experiments and so on? It is very difficult to deny that through the empirical case studies of EPOR, Collins has argued for a rather problematic compartmentalization between the scientific sphere and the social world.

EPOR: Backward and forward

To conclude, all the issues raised are not aimed at discrediting EPOR; quite the opposite. As with the work of Merton, it seems that there are certain aspects of this approach that should be retained, but may also benefit from further elaboration. Some key ideas are the following:

- The social dimension of scientific knowledge is undeniable, but its presence should *not* signify the relegation of natural, scientific and technological elements. The role these play in the advancement of knowledge should be recognized and theorized properly.
- Scientific debates are not free-floating events of a relativistic nature. Rather, they constitute conjunctures that—in democratic societies—unfold within the institutional and normative arrangements of the scientific field they belong to.
- The above point, however, does not imply that external influences from various institutional fields are not present. For the researcher to assess those properly, some form of *methodological bracketing* has to be applied that will help shift the focus of attention within and outside the laboratory or scientific community without recourse to any form of reductionism.

Actor-Network Theory

Actor-Network Theory, sometimes also known as the sociology of translation, was primarily developed by the Paris group of science and technology studies, and more precisely by Bruno Latour, Michel Callon, and John Law[2] (Latour 1996, 2). Presenting Actor-Network Theory in a concrete and convincing way can be an arduous task, as its proponents argue that ANT is not a theory (Callon 1999, 194). Law has further downplayed ANT's importance by claiming: "Like other approaches actor-network theory is not something in particular" (Law and Hassard 1999, 10). ANT resembles more of a fluid, elusive, ever-changing approach that has been applied to a great deal of issues ranging from life in the laboratory (Latour and Woolgar 1986) to museums (Hetherington 1999) and from aircraft (Law 2002) to Einstein's general theory of relativity (Latour 1988a).

While its proponents are rather adamant that ANT should not be construed as a theory, or that is "not something in particular," still there are some principles that ANT theorists espouse when researching the social world. Therefore, I will divide this section according to ANT's three main principles—as these were sketched by Callon in his seminal 1986 article, "Some Elements of a Sociology of Translation: Domestication of the Scallops and the Fishermen of St Brieuc Bay"—and extend the discussion by extensive reference to the works of Latour—who more than anyone has associated his name with this post-humanist approach—and Law. While it is true that ANT has received vigorous criticism mainly for its philosophical origins (i.e., relativism), its principle of symmetry (human and nonhumans are both actors) and its overall semiotic approach (Collins and Yearley 1992; Yearley 2005), it is also true that ANT "has been often misunderstood and hence much abused" (Latour 1996, 2). In my effort to offer an understanding of what can be kept from the sociology of translation, I will not repeat any criticisms that are targeted at ANT's philosophical principles and overall approach, but turn my attention to the way ANT uses key sociological notions such as actor, network, and structure.

Agnosticism

One major problem that ANT theorists discern in many studies of science and technology is that very often researchers tend to present their sociological account in a very partial and subjective manner. "The sociologist tends to censor selectively the actors when they speak of themselves, their allies, their adversaries, or social backgrounds" (Callon 1986, 98). The problem with such accounts is that social scientists tend to project their own preconceptions and assumptions on situated agents and technologies without demonstrating the necessary empirical evidence that justifies these claims. In doing so, they end up espousing some form of reductionism. One such form of reductionism, for example, is the argument that "*either* machines *or* human relations are determinate in the last instance: that one drives the other" (Law 1992, 382, stress in original).

For ANT, however, the answer does not lie in some meso-level or co-productionist approach. Rather, ANT theorists argue that reductionism can be only avoided with *agnosticism*. Agnosticism in ANT calls for the researcher to be "impartial towards scientific and technological arguments," for example, and also to "abstain from censoring actors when they speak about themselves or the social environment." In effect, by abandoning all preconceptions "no point of view should be privileged" and no interpretation suppressed (Callon 1986, 198–99). ANT is correct in arguing that

> there is no reason to assume, *a priori*, that *either* objects *or* people in general determine the character of social change or stability. To be sure, in particular cases, social relations may shape machines, or machine relations shape their social counterparts. But this is an empirical question, and usually matters are more complex. (Law, 1992, 383, stress in original)

ANT's first principle, however, entails much more than a simple rejection of deterministic or reductionist reasoning. ANT's agnosticism is aimed at rejecting *all* prior knowledge of the social world that the researcher may claim to have. Naturally, ANT's first principle is very suspicious of any theory of action or any prior assumptions regarding an actor's dispositions, motives, intentions and so forth. "ANT is based on no stable theory of the actor; rather it assumes the *radical indeterminacy* of the actor. For example, the actor's size, its psychological makeup, and the motivation behind its actions—none of these are predetermined" (Callon 1999, 181–82, stress added). Therefore, "instead of constantly predicting how an actor should behave, and which associations are allowed a priori, ANT makes no assumption at all, and in order to remain uncommitted needs to set its instrument by insisting on *infinite pliability and absolute freedom* (Latour 1996, 9, stress added).

Generalized symmetry

The second founding block of ANT is the principle of "generalized symmetry." The concept of symmetry forwarded by Bloor has been extended considerably and now argues that the narrative or repertoire used by the researcher should be one and the same for both technical and social aspects of the problem studied (Callon 1986, 200). Therefore, the same grid of analysis and the same vocabulary have to be used whether the researcher is examining social, technological or natural aspects (Callon 1986, 212). This practice is demonstrated perfectly in Callon's seminal study in which, for instance: the three researchers "join forces with the scallops, the fishermen, and their colleagues" (1986, 203); or "the scallops, and the fishermen are on the side of the three researchers in an amphitheatre at the Oceanographic Centre of Brest one day in November 1974" (1986, 210); and, finally, "the larvae detach themselves from the researchers' project and a crowd of other actors carry them away. The scallops become dissidents. The larvae which complied are betrayed by those they were thought to represent. The situation is identical to that of the rank and file which greets the results of Union negotiations with silent indignation: representivity is brought into question" (1986, 211). These examples from Callon's work demonstrate how the abandonment of all preconceptions at the theoretical level are reflected in the actual empirical project. For ANT, the researcher has to place the social, the natural and the technological on the same surface and examine them on exactly the same terms; doing otherwise would disclose, as it seems, some form of eclecticism or preconception.

The end of dualisms and the advent of free association

The previous two principles have paved the way for the third, and final, tenet of ANT, which is the principle of "free association." This principle actually signifies ANT's rejection of all the key dichotomies that seem to puzzle sociology. For proponents of ANT, dualisms such as human/nonhuman, society/technology and so on should not be upheld. For Latour, "every human interaction is sociotechnical [...] We are sociotechnical animals. We are never limited to social ties. We are never faced with objects" (1994, 806). In the same vein, Law argues: "If human beings form a social network it is not because they interact with other human beings. It is because they interact with human beings and endless other materials too [...] if these materials were to disappear then so too would what we sometimes call the social order" (1992, 382). For Callon, free association entails the following: "The observer must abandon all a priori distinctions between natural and social events. He must reject the hypothesis of a definite boundary which separates the two [...] Instead of imposing a pre-established grid of analysis upon these, the observer follows the actors in order to identify the manner in which these define and associate the different elements by which they build and explain their world, whether it be social or natural" (1986, 200).

Having mentioned ANT's three main principles, we can now turn to assess how certain key notions in this approach's repertoire are used. Most hopefully a potential initial reluctance to follow ANT's rather unorthodox vocabulary can be alleviated by keeping in mind that the sociology of translation is couched in the tenets of agnosticism (i.e., we know nothing about the social, natural, technological world), generalized symmetry (i.e., social, natural and technological events should be followed using the same grid of analysis), and free association (i.e., there are no clear-cut distinctions between society/technology, nature/

politics, actors/structures and so on, but only hybrids). Still, it remains to be seen whether ANT's hybrid vocabulary is, in reality, a step forward to understanding better social, natural, and technical relations.

Actants and actor-networks

As we have seen, the tendency to reject commonly held boundaries lies at the very core of ANT theory. In effect, for the followers of ANT, actors are not, and should not, be distinguished in terms of human and nonhuman. Instead, there should be symmetry between humans and nonhumans, as the two are so tightly woven into the endless fabric of networks that a distinction is virtually impossible (Latour 1994, 807).

In order to do away with the "traditional" term "actor"—which inevitably alludes to the human/nonhuman distinction—ANT theorists suggest that it is much more suitable to use the term *actant*; a term that comes from the study of literature. In a network, Louis Pasteur is an actant, and so is a machine (Latour 1988b). There is nothing to suggest, the sociology of translation argues, that an individual, a corporate body or a structural trait qualify as more or less "realist," "concrete," "abstract" or "artificial" actants (Latour 2005, 54).

Apart from their symmetrical reference to both humans and nonhumans, actants are also characterized by an unequivocally *relational* and *uncertain* nature (Law 1999, 4). So, what "lies at the heart of actor-network theory [...] is [the suggestion] that society, organizations, agents and machines are all *effects* generated in patterned networks of diverse (not simply human) materials" (Law 1992, 380, stress in original). But for ANT, uncertainty covers not only the form, qualities and functions of actants, but also the researcher's knowledge of who or what is performing the action. "To use the word 'actor' means that it's never clear who and what is acting when we act since an actor on stage is never alone in acting. [...] If an actor is said to be *actor*-network, it is first of all to underline that it represents the major source of uncertainty about the origin of action" (Latour 2005, 46, stress in original).

For the sociology of translation, actants do not have an ontology of their own, but their very existence, their actions, their functions and so on are only *outcomes* or *consequences* of their relations with other actants. Actants and networks cannot be separated ontological or analytically. "Actors are network effects" (Law 1999, 5). By being networks effects, actants only exist when they make some sort of impact. "An invisible agency that makes no difference, produces no transformation, leaves no trace, and enters in no account is not an agency. Period. Either it does something or it does not" (Latour 2005, 53).

Further clarification regarding the ontology of actors can be found in Latour's sociological assessment of Einstein's general theory of relativity, in which the French sociologist argues that a "giant in a story is not a bigger character than a dwarf. It just does different things [...] *Size is not a property of characters, only of networks and their relations* (1988, 30, stress in original). By bringing a giant and a dwarf on a par, Latour seems to be claiming that there is no ontological difference between the two. For ANT, this is a perfectly legitimate assumption, as Latour elsewhere overtly claims that "we have to try to keep the social domain completely *flat*" (2005, 171, stress in original). What matters is not size and zoom (i.e., macro/meso/micro scales), but the connectedness of actor-networks (Latour 2005, 187). But what are these networks?

For ANT, networks are not synonymous with *social* networks that study the social relations of human actors in terms of their distribution, frequency, homogeneity, the strength of their ties and so on. "[ANT] does not wish to add social networks to social theory but to rebuild social theory out of networks. It is as much an ontology or a metaphysics, as a sociology" (Latour 1996, 2). "Network is an expression to check how much energy, movement, and specificity our own reports are able to capture. Network is a concept, not a thing out there" (Latour 2005, 131). What networks do is "summing up interactions through various kinds of devices, inscriptions, forms and formulae into a very local, very practical, very tiny locus" (Latour 1999, 17). Nonetheless, networks do not exist on their own in ANT. The term is inextricably linked to that of the actor; "an actor is also, always, a network" (Law 1992, 384).

Finally, while nothing is known about actants in advance since, as we have seen, they are effects or consequences of the actor-networks they belong to, there appear to be some virtual entities that actually pre-exist actants. "Any given interaction seems to *overflow* with elements which are already in the situation coming from some other *time*, some other *place*, and generated by some other *agency* [...] There is indeed in every interaction a dotted line that leads to some virtual, total, and always preexisting entity" (Latour 2005, 166, stress in original).

Despite his conviction that the actor/structure debate constitutes "an artificial question" (Latour 2005, 169), Latour is obviously alluding to some structural activity. This becomes evident when he refers to them directly and discusses, though in passing, their ambivalent properties. For Latour, "Structure is very powerful and yet much too weak and remote to have any efficacy" (2005, 168). For the French sociologist, it is also very clear that structures "are not oozing out of a global context, of an overarching framework, of a deep structure" (2005, 193). Therefore, instead of trying to find structural entities stemming from a global context or deep structure, ANT researchers should focus on a flat ontology and "follow the actors themselves» and the heterogeneous actor-networks they form" (Latour 2005, 179).

ANT: Backward and forward

There is no doubt that the reservations ANT theorists have expressed toward some common practices of sociologists in SST are very well founded. In other words, it is true that assuming a priori actors' motives, intentions and so on without presenting the according empirical evidence does not actually add rigor to one's account. The same holds true for instances when social or technological forces are ascribed with a causal or determining influence in an ad hoc manner. While ANT is indeed right in pointing to some weaknesses of certain SST studies, it is not equally sure that the alternative courses of action they advocate are free of problems.

The first point has to do with the idea of *agnosticism*. As we have already seen, ANT theorists call on the researcher to abandon all preconceptions and all prior knowledge regarding the social, natural and technological elements that appear in one's study. My objection to agnosticism is not so much related to its implication (Is it really that beneficial to bracket all knowledge previously acquired?), but to the feasibility of its application and, more specifically, to its implementation by ANT's own advocates. When John Law says, "I assume that the world is materially heterogeneous, a mix of the social, economic, material, human, 'natural,' and technical" (2013), is he not bringing forward his own assumptions and pre-conceptions? When Bruno Latour, after a certain point in his academic life, started using *de facto* in his studies the hybrid vocabulary of ANT, did he actually espouse the agnosticism that the sociology of translation so fervently evangelizes, or did he use theoretical tools that were products of previous studies?

The second point has to do with the principle of *generalized symmetry*. To reiterate: for ANT, the same kind of vocabulary and analysis should be performed for the social, natural and technical aspects of a research question. As was already demonstrated, Callon strives to remain faithful to this tenet by placing fishermen and scallops on a par (after all, they are both *actants* for ANT) and by giving them exactly the same significance in the unfolding of events. For Collins and Yearley, however, this symmetry is only achieved at face value. This is so because in the very first place the creation of symmetry lies in the hands of the analyst. It is the social scientist who is in control the whole time and imposes this superficial symmetry on his narration. The rhetorical questions of Collins and Yearley entail both humor and, it seems to me, irony toward Callon and Latour but, nonetheless, they do make a trenchant point regarding the feasibility of ANT's second principle:

"Would not complete symmetry require an account from the point of view of the scallops? Would it be sensible to think of the scallops enrolling the scallop researchers so as to give themselves a better home and to protect their species from the ravages of the fishermen? Does the fact that there is no *Sociological Review Monograph* series written by and for scallops make a difference to the symmetry of the story?" (Collins and Yearley 1992, 313).

The third point has to do with the principle of *free association* and the forging of a hybrid vocabulary. By arguing that the distinctions of micro/macro, society/nature/technology, agency/structure, human/nonhuman and so forth are nonexistent and should be ignored, ANT theorists suggest that a new repertoire should be built in order to describe the world as they see it. But here lies a solemn problem. Ted Benton and Ian Craib point out this logical fallacy in no ambivalent terms:

"For one thing, Latour's key concepts for defining 'actants' are those of 'hybridity,' 'quasi-object,' and so on. These terms get such meaning as they have only in terms of the *prior* understanding of what 'subjects,' 'objects' and the 'pure' elements of the 'hybrid' are. Latour contravenes his own methodology in the very act of defining his most basic ideas" (2001, 71).

This disturbing realization that ANT's decision to depart from the "traditional" sociological vocabulary of actor/structure, micro/macro and so on is simply an untenable task also becomes evident when these terms make their appearance from time to time in ANT's narrative. If these notions do not exist why is it that Latour theorizes with them now and then? What appears to be a fundamental problem for ANT, therefore, is that the theorists who have formulated this approach tend to violate the approach's foundation blocks, which they have laid out themselves.

Moving from ANT's principles to its key terminology, we can discuss in greater detail whether Callon's, Latour's and Law's decision to shift to an altogether different vocabulary actually enhances the approach's heuristic value. ANT's main ontological and methodological claims about actants and structures can be summarized as follows:

- Following a semiotic approach, human and nonhuman actors (i.e., actants) are indistinguishable;
- Actants do *not* have varying size;
- Actants acquire their form as a *consequence* of the relations within which they are located;
- Actants are both *means and outcomes* of a practice;
- An actant's size, motivation and psychological character are *not* known in advance;
- Structures are *both* powerful *and* weak;
- Structures do *not* emerge out of a global context or deep construct;
- All of the above should be placed within a conception of a society that is *flat*.

To reiterate, ANT is right in arguing that knowledge of actors' motives, size, intention and so on should not be surmised without proper empirical evidence. ANT is also right in urging the researcher to *follow* how actants are related to one another and, instead of staying at the level of abstract ontology, to place emphasis on the ontology-in-situ and to ask *specific* questions that require *specific* answers. Nonetheless, in its endeavor to reject falsely projected knowledge and abstract armchair theorizing, it seems that ANT has ended up being in a state of complete denial regarding key aspects of the social, natural or technological world, so to speak. Such quasi-apophatic ontology has, at least, four problematic aspects.

First, by arguing that actors do *not* have a varying size, ANT is actually rejecting the fact that actors contribute *unevenly* in the production of everyday social reality. By taking the previous example of Latour—that a giant and a dwarf do not have a different size—can one really argue that, within the enormous and complicated network of agbiotech, the CEO of Syngenta and the director of a small laboratory have the same ontological size? Is there not a huge difference of resources they have at their disposal? Is that not significant, or is their "connectedness" as *actor*-networks all that matters? It seems that ANT's commitment to a flat ontology that can stretch wide and capture the dynamics of actor-networks has a major drawback: namely, it cannot accommodate social hierarchies. By not providing for the existence of hierarchies, the sociology of translation is simply in denial of the very simple fact that social actors contribute unevenly to the production, reproduction and transformation of social life.

Second, it *can* be the case that the researcher knows in advance *certain* things about actors. Latour himself acknowledges the presence of some "dotted lines that lead to a virtual preexisting entity" and I can see no reason the researcher cannot claim to have a modest degree of knowledge of these preexisting entities. Each social actor enters an interaction with a given set of deeply seated dispositions—what Bourdieu (1977) calls *habitus*; that is, a set of internalized cultural elements acquired through prior socialization—and a given role/position. Depending on the depth of research conducted, the researcher may have *some* knowledge regarding the actor's dispositions and role. For example, by carrying out interviews the researcher may know that situated agent X is of strong liberal convictions, highly values hard work and is a fervent opponent of state intervention. Furthermore, the researcher may already know that the specific situated agent became a junior manager at the company he works for, has a *spe cific* number of people directly under his authority, will get exactly *that* bonus if he achieves the goals set by the board of directors and so on. Having said this, there is no doubt that as time passes by and interactions unfold many things can change. My argument here is that it is one thing to censor actors so that the researcher's preconceptions are "miraculously" verified and quite another to reach the other extreme and argue that *nothing* is known in advance. It is true that how social interactions will eventually unfold should be open to empirical investigation, but it is equally true that empirical evidence may offer the researcher *some* insight into the situated agent's beliefs, intentions and structural positions.

Thirdly, it seems that ANT is underestimating the internal powers and active agency of actors when it argues that actants are merely the *effects* of networks. Should not one also look at the emergent characteristics, capabilities and susceptibilities of situated actors? Also, within the structural environment in which a situated agent is found, is there not some smaller or greater space for the creativity of action? Why is it that Latour expresses his ideas in that specific way? How is it that he has chosen those particular topics for his sociological research? Does not it all this have to do, to some extent, with his own decisions, with the ways that Latour managed the opportunities and limitations he was presented with? Would it, therefore, be just and accurate to claim that Latour is just a network effect? In essence, ANT is rejecting the rapprochement and acknowledgment that has been achieved in STS between realist and constructivist schools that ontological entities—with their own embodied powers—both *preexist* the situation and are also constructed at the same time (Heur, Leydesdorff and Wyatt 2012: 357). By arguing that the ontological status of the entities involved should be construed as an "accomplishment" (Woolgar and Lezaun 2013), ANT theorists tend to efface the rich ontological layer of actors with their intrinsic causal powers, knowledge, dispositions and so on. If the adequacy of an ontology is to be judged in terms of its explanatory effectiveness and the range of questions it allows one to address with relative success (Stones 1996, 28), I am very doubtful that certain of ANT's ontological claims are a step in the right direction.

Finally, this leads us to the limited way in which structures are presented: that is, that they are both powerful and weak. It seems it is much more suitable to argue that structures per se are neither powerful nor weak. Their strength or weakness is always related to that of an actor. We have to examine *empirically* how an actor is situated in relation to a structure and then assess the possible trajectories that an action may take. Some actors, such as the government's CSA, have the ability to affect structural conditions in multiple and profound ways, while others can do nothing more than take the structural environment they are faced with in a taken-for-granted manner.

The hybrid vocabulary forged by ANT theorists does not appear to offer a convincing way out of some of the problems that Callon, Latour and Law have quite rightly drawn attention to. Even if one is willing to overlook the self-contradictory nature of ANT's principles, how actants, networks and structures are theorized does not help the researcher grasp the rich and complex ontology of the social world. Nonetheless, ANT's suggestions that the researcher should "follow actors around" and be able to answer specific questions about specific agents and technologies (i.e., focus on the level of ontology-in-situ) are invaluable and should be seriously taken into account.

Andrew Pickering and the Mangle of Practice

The fourth and final contribution to the field of STS to be examined in this chapter is the work of Andrew Pickering. As mentioned in the introduction, Pickering's case is quite interesting in that, while he started his work as a disciple of EPOR, he soon began to deviate from the premises of empirical relativism. This departure from EPOR culminated in his most recent works (1993, 1995), in which a new approach called the *Mangle of Practice* was formulated. This approach reveals an affinity to some of ANT's principles, but also boasts some significant differences. Pickering's contribution, therefore, can be seen as a creative articulation of ideas stemming from both EPOR and ANT. In order to avoid needless repetition of points already alluded to, I will not juxtapose the Mangle with EPOR and ANT, but will primarily focus on the fresh ideas introduced by Pickering and examine how useful they can prove to be.

Pickering's initial empirical case studies (1981a, 1981b) offered service to the approach of EPOR. His work on the magnetic monopole (1981) was in fact heralded by Collins as another fine example of how scientific debates are closed (Collins 1983, 274). At the time, alongside Collins and other EPOR writers, Pickering was arguing for the presence of a social element in the advancements of science and technology. He argued that the debate concerning scientific observation can be distinguished on two levels: the "instrumental" and the "phenomenal." The instrumental refers to debates that center on the apparatus, the techniques and procedures that offer scientists a set of data, while the phenomenal refers to the discussions concerning the interpretation of those data (Pickering 1981a, 65). In "The Hunting of the Quark," Pickering recapitulated his belief in the social facet of science by stating that "experimenters have to *argue* for acceptance of their results within their community, and when they do, the instrumental and phenomenal images are part of the currency of exchange. Experiments are performed and evaluated within a socially sustained matrix of commitments—beliefs and practices—and this matrix severely constrains the acceptable constructions an experimenter can place upon his esoteric experience" (Pickering 1981b, 235, stress in original).

Pickering's work initially received outstanding reviews. His contribution was heralded as "excellent, incisive, and pioneering" (Heilbron 1986, 1479). Yves Gingras and Silvan Schweber (1986) found that "Pickering's exposition is rich in insights and reflects his command of the subject matter. Indicative of this mastery is the dearth of equations to be found in the book" (1986, 372). Almost a decade later, Pickering introduced his concept of the Mangle, a theoretical construct that has aspired to be much more than merely another approach in

SST; a construct that claims to be a "Theory of Everything" (Pickering 1995, 248). This time, however, reviews of Pickering's work were not so favorable. Gingras, for example, who once enthusiastically lauded Pickering's work on Quarks, was left puzzled by the Mangle's premises. In fact, the debate between the two (see Pickering 1999 and Gingras 1999) involved personal insults and ended with Gingras bitterly commenting that "[Pickering] is contemplating [his] own oeuvre from the top of a mountain, while looking down on those who tediously try to make sense of the bits and pieces of arguments collected in a book and who, finding them wanting, simply point to inadequacies, ready to be enlightened in their valley of the blind" (1999, 315). But what were the issues the sparked such a controversy?

Post-humanism in Pickering

The Mangle signifies Pickering's radical break from his earlier works in EPOR. In fact, traces of ANT are evident throughout the Mangle, and Pickering would declare himself "happy if the book were read as an attempt at constructive dialogue with the actor-network [theory]" (1995, 11). Nonetheless, Pickering's work is no simple reiteration of ANT.

The author of the *Mangle of Practice* argues for a post-humanist approach that displaces the "traditional" interpretive frameworks of STS that regard human individuals and groups as the locus of understanding and explanation (Pickering 1993, 561). Instead, Pickering stresses the "need to recognize that material agency is irreducible to human agency if we are to understand scientific practice" (1995, 53). While this post-humanist approach does not give primacy to human agency, it does not place humans and nonhumans on a par either. For Pickering, humans and nonhumans differ in that the former have *intentionality* (1993, 565). However, apart from intentionality, Pickering recognizes that the "scientific practice is typically organized around *specific plans and goals*" (1995, 17, stress in original). In scientific practice, however, scientists are not alone in their pursuit of their plans and objectives. Their actions unfold in relation to the materials they interact with and which, for Pickering, have their own *agency*. A key characteristic of *material agency* is its *temporal emergence*. "The contours of material agency are never decisively known in advance, scientists continually have to explore them in their work, problems always arise and have to be solved in the development of, say, new machines" (1993, 564). But for Pickering this also holds true for human agency: "Human agency is, just like material agency, temporally emergent. We can say more about the intentional structure of the former, but, in the end, it too simply emerges in the real time of practice" (Pickering 1993, 566). So, in Pickering's post-humanist approach, agency can be found in both humans and materials. Both are temporally emergent as they are "mutually and emergently productive of one another" (Pickering 1993, 567), but humans differ in their intentionality and ability to set goals.

What now happens in scientific practice is that human and material agents are engaged in the dialectic of resistance and accommodation; which is exactly what defines the Mangle of Practice (Pickering 1993, 567). The Mangle can also be visualized as a "dance of agency."

> The dance of agency, seen asymmetrically from the human end, thus takes the form of *a dialectic of resistance and accommodation*, where resistance denotes the failure to achieve an intended capture of agency in practice, and accommodation an active human strategy of response to resistance, which can include revisions to goals and intentions as well as to the material form of the machine in question and to the human frame of gestures and social relations that surround it. (Pickering, 1995, 22, stress in original)

In the postscript of his book, Pickering is, of course, more than content with the explanatory power of the Mangle. This dialectic of resistance and accommodation offers an answer to almost every question in sociology, history, biology, art and industry, just to mention a few areas. For Pickering, mangling processes are everywhere. "And so, in the end I allow myself to be overtaken by hubris in thinking of my analysis of scientific practice as a potential TOE, a theory of everything" (Pickering 1995, 248).

Hubris and the heuristic value in the Mangle: Backward and forward

Whether Pickering has allowed himself to be overtaken by hubris or not is unimportant. What is important, however, is his contribution to the field. In order to assess the significance of the Mangle in STS, I will revisit its key components in the order previously presented.

Pickering's post-humanism is couched in the belief that human-centered approaches are inadequate in portraying scientific advancements since materials, apparatuses, quarks, magnetic fields and natural phenomena are, in general, theorized as "dead" and passive recipients of human action. To make up for this deficiency, Pickering has argued that agency ought to be attributed to materials; giving us, therefore, a "material agency." This material agency is always *temporally emergent* and differs from the human agency in that it lacks intentionality and goal-setting.

While Pickering is right in pointing out that the tendency of humanist approaches like EPOR to downgrade the ontology of nonhumans is indeed a source of concern, the ascription of agency to materials is equally worrisome.

To begin with, the very choice of the term "agency" for materials can be confusing. In Pickering's work, agency for materials and objects is limited to acts of "resistance" and nothing more. In the works of scholars who have devoted lengthy passages theorizing agency—such as Alexander (1998), Archer (1996) or Stones (2005)—the concept takes an incomparably richer form. While these three theorists may not describe agency in exactly the same words or agree on all issues, at the very least there is most certainly a consensus that agency involves, among others, the abilities to be creative, to typify stocks of knowledge, to be inventive, to devise strategies for action and more. Even if one finds Pickering's use of agency to simply describe material resistance a perfectly suitable choice, a much more serious problem arises when the material agency is inextricably linked to the phenomenon of temporal emergence.

Throughout the Mangle, Pickering "insists that talking about resistance does not suggest that what resists, and the resistance itself is always there but, on the contrary, that it emerges *only* in the interaction with the scientist" (Gingras 1997, 322–23, stress in original). For Gingras, this insistence of Pickering on the emergence of material agency renders his proposition either trivial or metaphysical. It is trivial in the sense that *by definition* we can experience resistance only during the interaction; to repeat, Pickering has already clearly informed us that resistance exists "on the boundary of human and material agency" (Pickering 1995, 67). It can alternatively be seen as metaphysical in that Pickering leaves open—or in fact invites—the possibility of a *strong* emergence of agency or resistance. As Gingras notes "it is hard to understand how a 'block' can arise or emerge if it is not already there waiting to be 'seen,' as it were, in the interaction" (1997, 323). To my mind, further confusion is created when metaphysical or overall transcendental qualities are attributed to the mangle itself, which eventually dwarfs all human and nonhuman agencies. Agency appears to be bestowed to the Mangle when Pickering argues that "It is the mangle that determines in time what scientific machines will look like, what scientists will believe, how they will conduct themselves around those machines, and how (if at all) these pieces and others will hang together and relate to one another" (1995, 112).

Lynch is right in arguing that "it is highly misleading to suggest that a unified 'something' pervades and directs an open-ended field of actions, contingencies, determinants, and judgment" (Lynch 1996, 811). The ontology of the mangle and the ways in which it determines humans and nonhumans—which concurrently comprised the mangle in the first instance—remains concealed. Pickering's "presentation lacks clarity at key points because of his reliance upon metaphorical language in the place of detailed analysis" (Sargent 1998, 721).

It may have become apparent by now that, via the mangle, Pickering is trying to transcend the agency/structure dichotomy. In doing so, however, he theorizes neither agency sufficiently nor structural constraint properly. To repeat, for Pickering constraint is derived from social structure and has two chief characteristics: (a) it is "*nonemergent* [...] on the time scale of human practice" and (b) it is the "language of the prison" (Pickering 1995, 65, stress in original). I think both propositions are only partially correct. It is true that structural constraint can preexist social action and endure through it, but this is not necessarily the case. While this may be true for large historical forces or Durkheimian "social facts," such as capitalism, industrialization, anomie and so forth, it is not true for instances in which structures are formed as interactions unfold. For example, during a presentation on the progress of a research team, the head of the laboratory may decide to stop funding that particular project. Does not that constitute an *emergent* constraint? Such a decision had not been taken beforehand—it did not pre-exist, in other words—but was instantiated because the briefing brought unsatisfactory results to the surface. At the same time, structural presence is not necessarily prison language. Structure can appear both as constraint and as enablement. If a company, for instance, decides to restructure its organization by making advertising personnel redundant and concurrently hiring scientists for the R&D sector, the former group will see that as a constraint, but the latter will rejoice at the opportunity presented.

To conclude, Pickering has offered to the field invaluable insight regarding the social aspect of scientific advancements, and the sterling reviews his earlier works received attest to that claim. Also, his critique of traditional SSK that apparatuses and technologies should be construed as much more than dead material is certainly moving in the right direction. Nonetheless, his aspiration to explain almost everything under the sun as merely a process of "mangling" is at best inadequate. In the process of forging his own post-humanist approach, Pickering has forsaken his earlier work; the useful distinction between instrumental and phenomenal levels now plays a peripheral role, and the birth of scientific facts is explained by mere reference to the process of resistance and accommodation. Furthermore, the role of the scientific community has been ousted from the mangle without any apparent or convincing reason. It seems that Pickering's intention to radically break away from EPOR and ANT has, in the end, led him to invent a terminology that is all too conveniently succinct, but also so disturbingly unrevealing of anything.

Ontology and Methodology in STS: Toward a Critical Synthesis

In this chapter, four key approaches in STS were assessed: the Mertonian sociology of science; Collins's SSK; ANT in the writings of Latour, Callon, and Law; and Pickering's *Mangle of Practice*. It was argued that, on both ontological and methodological levels, each approach offers significant contributions to the field. Nonetheless, it was also demonstrated that there are certain aspects in each approach that would most likely benefit from minor revisions or lengthier elaboration, while other constructs proved of little heuristic value. An ontological and methodological map of the key approaches developed in this chapter, together with some subtle modifications that have arisen from the review, can be seen in Table 3.1. All of these germane points will be brought forward to the next chapter, where they will find their place within the suggested theoretical framework.

Table 3.1 Key contributions of the four approaches together with suggested modifications.

Theorist/ Approach	Ontology	Methodology
Merton	The scientific field is a sphere of relative autonomy with its own normative and institutional arrangements. Scientists ideally internalize and follow the field's norms, but this is not always the case. Other spheres with their own distinct logic and normative patterns may affect science and impose, to a smaller or greater extent, their own logic.	Scientific advancements or debates should *always* be investigated within their social context; that is, they should be appraised in tandem with the scientific community they were born in. At the same time, scientific and technological products can be appraised in accordance with the effect they have on other institutional spheres.
Collins/SSK	Scientific advancements have an undeniable social character. Social influence stemming from outside the scientific community is *always* present during scientific interactions.	The external influence should be *methodologically bracketed* so that the examination of different social locations is carried out in an orderly and analytically separate manner.
ANT	Nonhumans should *not* be construed as passive recipients of human action. Humans and nonhumans should *not* be compartmentalized, but should be positioned within the heterogeneous networks they form.	Actors should not be censored, and the researcher should not ascribe to them attributes that are not based on empirical evidence but simply reflect his own preconceptions. Furthermore, the researcher should "follow actors around" in his effort to map the social terrain and understand who and what is involved in the interactions that unfold.
Pickering	Materials should not be construed as passive recipients of human action since, at times, they exhibit some form of resistance to human actions.	One way of understanding scientific progress is by focusing on how scientists interact with apparatuses, machines, tools, genetic code and so on.

Chapter 4

BENTON, MOUZELIS, STONES: SOME KEY ADVANCES IN CONTEMPORARY SOCIOLOGY

At the center of the debate between the humanist and post-humanist approaches, which we examined in the previous chapter, lies the nature of the relationship between humans and nonhumans. As we saw, each of the three approaches in STS offers a different viewpoint. Collins's humanist EPOR focuses on human actors and ascribes to all nonhumans a sense of weakness and malleability; ANT's semiotic post-humanism argues that there is no ontological or theoretical difference between humans and nonhumans as they are all *actants*; Pickering's Mangle argues that the difference between human and nonhuman agents is that the former are capable of intentionality and goal-setting while the latter can only exhibit a temporally emergent resistance.

It has been argued so far that none of these viewpoints is satisfactory, as they advocate an ontology that either leads to reductionism or paints a very impoverished picture of the nonhuman world. These shortcomings are not simply a matter of theoretical interest but constitute limitations which may seriously impair our understanding of the social and natural worlds. The problematic ontology of these three schemes can be better understood by choosing examples from the field of agbiotech. The human/nonhuman distinction in EPOR, ANT and the Mangle appears to take the following form:

- For EPOR, a GM plant's inherent ability to repel aphids, reduce the use of pesticides or increase yields is not an objective ontological reality but a product of scientific negotiation and interpretation;
- For ANT, a field of GM canola and a union of Canadian farmers are all indistinguishable actants that simply "do different things" in the heterogeneous actor-network they form;

- For Pickering, GM wheat and a Rothamsted biologist are distinguishable only because the latter possesses the power of intentionality.

It becomes obvious, therefore, that an analysis of agbiotech cannot be realized so long as the human/nonhuman distinction is not properly addressed, initially on an ontological basis and, subsequently, on a theoretical one. In the next section, I will argue that this highly problematic area of STS can find a solution in the work of Ted Benton.

The Contribution of Benton

The multi-dimensional and diverse work of Benton traverses from the natural to the social world by offering insights to a variety of issues ranging from the behavior and ecology of bumblebees (2006) and grasshoppers (2012) to key philosophical issues in sociology (1977 and 2001), Marxism (1984, 1996), and the environment (1994 and 2007). Through the writings of Benton, these seemingly unrelated fields of study are brought closer together in ways that were previously hard to imagine; they do so by making links: for example, between bees as social animals and ideas from critical political economy, political sociology and law in thinking about the causes of bumblebee decline and prospects for reversing it (2012, viii). Although throughout his prolific career Benton has not written extensively on the construction of scientific and technological artifacts per se, I think it will soon become apparent that his theorizing on human and nonhuman relations can be applied to STS with very interesting outcomes. I focus on two areas of Benton's work: human and animal relations and the hierarchy of sciences.

Human and animal relations

Contrary to previous theorists who have attempted to tackle the society/nature dichotomy, either by favoring one over the other or by rejecting the dualism *tout court*, Benton argues that the dichotomy must be overcome by embedding social and natural entities on a continuum and examining the relations these develop with one another. This ontological naturalism in the work of Benton argues for a view that human nature and society should be situated firmly within "nature," broadly understood, rather than in opposition to it (Benton and Craib 2001, 183). The hallmark of Benton's naturalistic approach is the belief that "those things which only humans can do are generally to be understood as rooted in the specifically human ways of doing things which other animals also do" (Benton 1993, 48). In order for the embeddedness of humans in the natural environment to be sustained, Benton argues, the human/nature dualism must be convincingly displaced. Such displacement can be achieved only when the shared characteristics of humans and animals are stressed (Benton 1993, 46–47). In summary, such interrelated features are the following:

- *An organically limited lifespan: birth and death.* While these are shared features of humans and animals, in the human case they acquire a distinct significance. Among others, this significance is contingent on the metaphysical, moral and affective dimensions of loss.
- *Temporal phasing of organic growth, development and decline in individual life-spans.* These features are also shared in the lives of humans and animals since in many nonhuman social species changes in social position, relations and patterns of activity are related to their organic transitions. Nonetheless, what is distinctive in humans is the length of infantile dependence (in proportion to the total lifespan), which is fundamental to the acquisition of complex sociocultural skills and capacities.
- *Sexuality.* Sexual reproduction and social regulation of sexual activity is a third feature shared by humans and nonhuman animals. However, humans construe sexual activity and sexual difference through the prism of complex sociocultural and psychological interpretations.
- *Social cooperation in the meeting of organic needs.* In order to survive, human and nonhuman animals have to engage in some sort of activity with their environment, such as hunting prey or exhibiting territorial defense. In order to ensure a higher probability of success in these tasks, some form of social coordination is necessary for issues such as regulating the distribution of the catch among participating members of the group or communicating to facilitate and enhance coordination. In humans, the use of symbolic modes of regulation of social interaction and coordination is developed to a unique and extremely sophisticated extent as compared to other social species.
- *Stability of social order, and the integration of social groups.* Both humans and many nonhuman social species (especially primates) have the ability to develop strong affective bonds (not necessarily with close kin), associations of groups, hierarchies of dominance and divisions of labor and consumption and more. In humans, such features require a social life of long-term integrity and stability (Benton 1993, 53–54).

To be sure, by stressing the interrelated characteristics of human and nonhuman animals, and by theorizing the natural as an integral part of the social, Benton is not trying to bring natural (as opposed to social) reductionism in through the back door. On the contrary, in his seminal article "Biology and Social Science," he is very clear in giving due credit to the importance of social forces.

I think it is quite wrong to think of scientific (including social scientific) discourses as *merely* aspects or effects of the power-play of wider social or political interests. It is nonetheless true that no science is wholly immune to such historical forces, and major conceptual shifts in the sciences are commonly linked to struggles going on in their wider historical context. (1991, 1, stress in original)

The hierarchy of sciences

In fact, the avoidance of reductionism, which has so much plagued STS theories, is one of the key concerns in Benton's writings. Apart from his commitment to a fundamental ontological continuity, Benton has also strongly argued for a "realignment" of the natural and social sciences (Stones and Moog 2009, 12; Benton 1991, 26). The realignment that Benton suggests is an argument against both compartmentalization and reductionism. This is so because, on the one hand, links among disciplines are drawn and, on the other hand, lower-level sciences are not given primacy of any sort. To be more precise: for Benton, sciences can be hierarchized as follows (Figure 4.1).

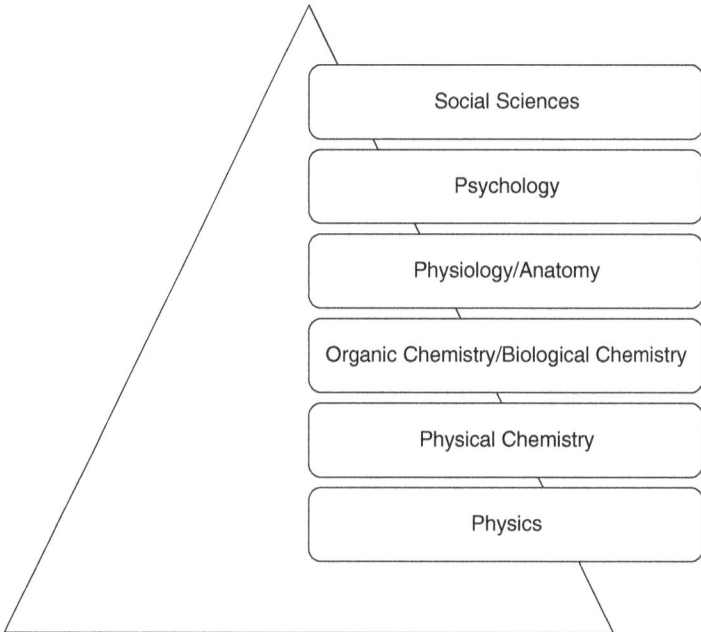

Figure 4.1 The hierarchy of sciences and the order of higher- and lower-end mechanisms.

This hierarchy of sciences suggests that each scientific discipline is concerned with a particular level of reality. The relation between sciences is threefold:

- The lower-level sciences explain (at best) the constitution of mechanisms at the higher level. This, however, leaves open questions as to when, where or how these mechanisms will be triggered. One example is that human beings have the anatomical and mental capacities to hunt animals in the wild and survive under adverse weather conditions. It is, however, not at all certain that all humans are ever going to use such capacities, or under what circumstances and for what purposes.
- Once higher-level mechanisms are formed their activities have effects on lower-level mechanisms. For example, a person who, for whatever reason, feels rejected by his immediate social environment may end up suffering from severe neuroses. By the same token, this severe stress or fear will take its toll on the central nervous and endocrine systems by altering the rate of chemical reactions involved in respiration and so forth. What follows from this point and the previous one is that causality can flow upward and downward in the hierarchy. What is also indisputable is that each level has its own specific reality that can *contribute* to explaining phenomena of the higher level, but never *fully* do so.
- Entities at higher levels have properties and powers that cannot be predicted in advance as a derivative of properties of lower-level entities. For example, a full genetic test can tell us a few things about a person's traits, genetic characteristics, disposition to illnesses and so forth, but it cannot say anything about the richness of that person's character, his emotional world, memories, aspirations and so on (Benton and Craib 2001, 126–27).

Now, if one is willing to accept this hierarchy of sciences and the relations among higher-level and lower-level mechanisms and examine in this light the writings of EPOR, ANT and the Mangle, some criticisms already raised can now be approached from a different perspective:

- EPOR has succumbed to a form of "reverse reductionism," if we are allowed to call it so, as it has retracted all mechanisms of physics, chemistry, anatomy and psychology to the domain of social sciences by espousing a top-down causality. Furthermore, this reductionism of mechanisms has also been extended to a reductionism of sciences, with Collins refusing to examine each phenomenon at a separate level but, more or less, imposing a social facet to every layer in the hierarchy;

- ANT has conflated all sciences and mechanisms by placing them on a flat ontology where semiotics and the principle of symmetry negate the possibility of a diverse ontology;
- In the Mangle, Pickering has tried to "enhance" the position of lower-end mechanisms by resorting to anthropomorphism and lending characteristics of one layer to another in an *ad hoc* manner.

Conclusion: From anathema to exegesis

The hierarchy of sciences, together with the human and animal relations discussed previously, helps us understand the rich, complex and interrelated ontology in the work of Ted Benton. It has been demonstrated that instead of transcending the human/nonhuman dualism by resorting to forms of reductionism or by espousing conflationist techniques, the British sociologist argues that the divide should be bridged by placing humans and nonhumans on a continuum.

> So, we are essentially social beings through our embodiment, not just our mental lives. But in noting this, we are also committed to an understanding of the relationships which constitute society as binding embodied humans to other living and non-living beings: to physical spaces, raw materials, tools and machines, domesticated and wild animals and plants, agricultural and semi-natural ecosystems, buildings, highways and so on. All of these things and relationships are produced, reproduced or transformed as elements in the overall metabolism of society. It follows that society cannot reasonably be represented as a single level in the hierarchy. (Benton and Craib 2001, 128)

While the human/nonhuman dichotomy has proved to be some sort of anathema for STS theorists, for Benton it has been a point of extensive and detailed articulation. I am convinced that this way of construing the social, natural, and material world is a very productive one as it helps release the tension that lies at the core of many STS approaches by offering a balanced, internally coherent and differentiated approach.

This holistic logic and notion of embeddedness in Benton's writings will underlie the overall orientation of the theoretical framework, which will gradually start to unfold in the following section. The theoretical construct endorsed is a combination of two seemingly incongruent approaches. On the more abstract ontological level, I use Mouzelis's (1990) post-Marxist Technology, Appropriation, Ideology scheme (TAI) and on the ontic level I use Stones's

(2005, 2010) Strong Structuration Theory (SST). The purpose of the section is, therefore, to familiarize the reader with the two theorists' contributions to contemporary sociological theory and underline the heuristic value that their propositions have to offer to the discipline at large and the sub-field of STS in particular.

The Contribution of Mouzelis

Post-Marxist alternatives

While most of Mouzelis's contribution to the discipline of Sociology is quite widely acknowledged,[1] his work on post-Marxism has remained neglected for the most part. Mouzelis himself has not at all disregarded his TAI framework and has been consistently restating the basic theoretical tenets that appeared in the *Post-Marxist Alternatives* throughout his later works (1994, 1995, 2008). Nonetheless, despite the Greek sociologist's constant articulation of TAI's principles and the favorable scholarly reviews,[2] there has been no apparent adoption of TAI by other theorists, neither on a theoretical nor an empirical level. But, what is actually argued in this post-Marxist scheme?

Holistic logic in Marxism

For Mouzelis, one of the aspects of Marx's theorizing that has to be safeguarded at all costs is its holistic logic. Drawing from Lockwood's (1992) fundamental distinction between "social and system integration," Mouzelis argues that one of the greatest assets of Marxism is that it allows the examination of both systemic or institutional incompatibilities (e.g., between forces and relations of production) and the ways that these incompatibilities lead or fail to result in class conflict, collective protest, development of class consciousness and so on (1990, 23). This means that one is allowed to examine societies in terms of both agentic/causal and systemic/functionalist perspectives since Marx's analysis provides theoretical links between system and agent. "Insofar as Marxism views the economy as an articulation of modes of production and insofar as the relations of production constitute the major feature of every mode, this key concept provides a bridge between a systemic/institutional and an agency/action approach" (1990, 25). Consequently, there is in Marxism, at least potentially, a balanced approach in the structural positions—social action interplay. An analysis of how the division of labor in society places actors in certain positions within the sphere of production and how this leads or fails to lead to social struggle, for example, can be conducted in a non ad hoc manner.

While Marxism is capable of avoiding actor or system essentialism, this does not mean that it can very easily avoid economic reductionism.

The relative autonomy of the political sphere and the base/superstructure dichotomy

For Mouzelis, even when Marxist theorists try to overcome economic reductionism—in other words when they refrain from assessing political phenomena by resorting to explanations pertinent to the capitalist mode of production or the conditions of existence of the capitalist economy—they do so in an unsatisfactory way.

A key concept in Marxist thought is the base/superstructure dichotomy, which, to put it briefly, "propound[s] the idea that the economic structure of society (the base) conditions the existence and forms of the state and social consciousness (the superstructure)" (Bottomore 1991, 45). In this sense, those who dominate the economic system have the ability to shape the superstructure according to their interests. In the base → superstructure model there is little doubt that the economy is given primacy but, nonetheless, the relation is reciprocal[3] (Outhwaite 1993, 43–44).

In an attempt to break away from any deterministic accusations that could be targeted toward the base/superstructure dichotomy, Althusser (1969) and fellow structuralists have ascribed to other institutional spheres a sense of "relative autonomy." In that case, "the economy is always determinant in the last instance but does not always play the dominant role; it may determine that either of the two superstructural levels be dominant for a certain period of time" (Bottomore 1991, 46). Or, in other words, "the economic is determinant in the last instance, not because the other instances are its epiphenomena, but because it determines which practice is dominant" (Storey 2006, 56).

Although Althusser does advocate that "the lonely hour of the last instance never comes" (1969, 113), nonetheless, for Mouzelis the problem remains as the other spheres, which have been granted a "relative autonomy," are still explained in terms of class struggles, change in mode of production and so on (1990, 13–14). For Mouzelis, therefore, the main problem in Marxist theory is that even if a "relative autonomy" is granted to the other institutional spheres, or even if they are ontologically construed as different from the economy, so long as there are no conceptual tools that enable an analysis of the polity or culture in their own right, economistic explanations of non-economic aspects are simply unavoidable. As a result, it is not sufficient to accept a "relative autonomy" of the political sphere on the theoretical level and at the same time to pay lip service to the methodological, but to examine social fields as relatively autonomous, both in theory and in practice (Mouzelis 1990, 50). But what conceptual tools are there that enable such a smooth transition between actors and structures within and across institutional spheres?

Technology, Appropriation, Ideology (TAI)

If doing away with the base/superstructure dichotomy is one thing that Mouzelis argues for, the other issue he proposes is that the concepts of technology, appropriation and ideology are aspects of not only the economic sphere but of the polity and culture as well. For him, "*the technological, the appropriative, and the ideological should be considered as constitutive elements of all major institutional spheres*" (1990, 79; stress in the original). Although the abolition of the base/superstructure concept may strike one as a significant point of departure from traditional Marxism, the placement of technology, appropriation and ideology at the center of analysis helps preserve theoretical bridges between Marxism and Mouzelis's suggestions, as the three constitute fundamental Marxist notions. What Mouzelis is particularly interested in retaining in his post-Marxist scheme is the balanced social/system integration approach, on the one hand and, on the other, the critical perspective that Marxism has at its core by so vividly portraying societies as arenas of potential struggle and exploitation. But what does Mouzelis mean by technology, appropriation and ideology?

Technology

If "technology" is to qualify as a constitutive element of all institutional spheres and social wholes, then there is no doubt that the term cannot be related only to technological artifacts, but has to be used in a much broader sense. For the author of *Post-Marxist Alternatives*, "technology" refers to the "technological means (means of social construction) through which men and women more or less unintentionally construct, reproduce, and transform their social existence" (1990, 79). Therefore, in the economic sphere, according to this syllogism, technology (or "technologies of production") is synonymous with forces of production (1990, 50). Accordingly, in the political sphere, one can find corresponding technologies; that is, forces of domination. These entail means of administration (administrative or military apparatuses, techniques of taxation, of national accounting, of mass mobilization and so forth) and means of destruction (1990, 51).

Although productive forces play a fundamental role in Marxist thinking, and there is controversy regarding the exact nature and qualities of this term,[4] Mouzelis does not offer a critical appraisal of the debate, but chooses to tentatively accept G. A. Cohen's (1978) interpretation of Marx's writings where productive forces include instruments, labor power and knowledge (1990, 51, 55). For Cohen (1978) a force or power is not a relation. "It is not something which holds between objects, but rather a property of an object, or, in an extended use that Marx indulges, an object bearing that property, an object having productive power, and such an object is also not a relation" (28). Mouzelis, too, leaves no room for social relations to appear in the discussion of technology, but adds to Cohen's definition that productive forces may also include " 'technical' organisational work arrangements," while at the same time stressing the materiality that both economic and political forces or technologies of production share (1990, 55).

Consequently, for Mouzelis technology is used in such a way that it encompasses both technological artifacts and institutional arrangements. For example, if one were to assess the technological aspect in the sphere of agbiotech, that is the means of agbiotech production, one could start by looking at the procedures and standards of available GM techniques, the required equipment that enables scientists to alter a plant's genetic sequence, the capacity of various laboratories and the tests and procedures they can carry out, the way research centers or relevant university departments are funded, the qualifications of scientists working in agbiotech, how their performance is appraised and so on. It seems to me that social actors are clearly involved in most of the instances mentioned, but Mouzelis has avoided any discussion of humans and their labor power. Stripping technology of any social arrangements is a move that encourages precisely the imbalanced portrayal of productive forces that Mouzelis has so much wanted to avoid. I discuss this issue in the following chapter. At this stage of analysis, the focus is on impersonal institutional settings; "who" questions are feasible when the discussion moves to the appropriative dimension of an institutional sphere or social whole.

Appropriation

The appropriative element of a field refers to the "patterned ways in which the[se] means of social construction are appropriated/controlled" (Mouzelis 1990, 80). If, for Mouzelis, technology refers to the means of (economic, political or cultural) production, then appropriation is synonymous with the corresponding relations of production. In the political sphere, for example, appropriation—or relations of domination—is defined as the "institutionalized ways of regulating the political division of labor, the distribution of political power between dominant and dominated groups" (1990, 66). The notion of appropriation is regarded as very significant since, according to the author of *Post-Marxist Alternatives*, it safeguards one of the most important comparative advantages Marxism has, at least potentially, over other theoretical paradigms such as conflict theory or normative function-alism: namely, the examination of social wholes in terms of both social and system integration.

With the notion of appropriation, actors are allowed to enter the analy-sis, and this enables the researcher to investigate "who" controls the forces of (social) production and in what ways is this achieved. Going back to the previous example, the appropriative aspect of the agbiotech sphere could refer to the actors (individual or collective) who are responsible for the design of risk-assessment frameworks, the allocation of funds for research and development, the legislation that allows or prohibits the commerciali-zation of GMOs, the protocol that needs to be followed in order for a field trial to take place and so forth. Such actors can include ministers and their consultants, associations of scientists, independent think-tanks, farmers' unions, members from the industrial sector and more. What should also be examined is via which patterned means do these actors have the authority to exert influence or wholly control the decision-making process. Is this done through institutionalized political processes? Is it a matter of scientific consensus? Or is it a matter of emergent circumstances that have arisen out of social conflict? Are the relations between the government and the associ-ation of farmers or the agbiotech industry orderly or conflictual? While this approach focuses on actors, Mouzelis, borrowing from David Lockwood, would argue that a comprehensive analysis needs the examination of sys-tem integration as well. In other words, one should also consider whether there are any contradictions between the functional prerequisites of parts of the system. Is there a contradiction between the needs of the agricul-tural sector and the needs of the economy? Is there a conflict between the norms of the scientific field and the political principles of democracy and openness to the public?

Ideology

For Mouzelis, ideology refers to the discursive and non-discursive methods that: "(a) give a distorted picture of how the means of production are owned/controlled and (b) provide reasons to justify and legitimize the prevailing con-trol/ownership arrangements" (1990, 68). This process of distortion or justi fication refers to the concealment of contradictions that may exist during the appropriation of the means of (social) production. What has to be made clear once again is that these discursive and non-discursive methods of justifying or distorting specific social arrangements exist in all institutional spheres (polity, education, religion and so forth). "As a matter of fact, insofar as all institutional spheres are partly constructed and reproduced by technologies, and insofar as such technologies are controlled by privileged groups, conditions will be such as to favor the emergence of mystificatory, justificatory practices" (1990, 71). Therefore, following the demise of the base/superstructure distinction, ideology is no longer seen as a "residual category" or a "blanket term" that belongs to the superstructure and whose function is to mystify or justify the economic relations of production. Within the TAI scheme, ideology is an essential component of all institutional spheres and, therefore, the possibility opens up that educational or religious institutions, for example, may have a logic and historical dynamic of their own (1990, 70).

The ideological methods of concealing can be related not only to covering structural inequalities (unequal distribution of wealth, political clientelism, non-inclusive education and so on) but also "to neutralizing antagonistic interests by presenting them as mere differences" (1990, 68). The Marxist concept of ideology, Mouzelis argues, does not merely refer to a set of abstract ideas, but entails institutionalized practices related to social locations within production and also socializing processes that render the subject more willing to acquiesce in the status quo. It has to be stressed that it is not only the exploited, but also the exploiting classes that may be unaware of the distorting and justificatory connotations that these social practices involve. Despite the all-pervasiveness of ideological themes in the various locations of social life, there are still instances in which these motifs can be refuted. "In certain conditions (for instance, when contradictions between forces and relations of production keep growing) the underprivileged may escape from the grip of the dominant ideology and develop counter-ideologies or utopias, which, by projecting what the future 'ought to be', challenge the status quo and undermine the 'obvious rightness' of the dominant classes' ideological themes" (1990, 69).

Having offered an introductory overview of Mouzelis's post-Marxist framework, we can now turn to discuss Rob Stones's Strong Structuration Theory and its significance for contemporary sociological theory.

The Contribution of Stones

Strong Structuration Theory

Rob Stones's seminal work, *Structuration Theory*, is a theoretical framework that refines, develops and enriches the core concepts of Giddens's homonymous theory. The outcome of Stones's fruitful criticisms of Giddens is the Strong Structuration Theory (SST), a theoretical framework that remains true to Giddens's central idea of the *duality of structure*,[5] but nonetheless departs from the initial ontological and methodological premises in substantial ways. As a response to Giddens's neglect of epistemology and methodology, Stones takes the vague and indeterminate ideas of ST and embeds them in the realities of the social world (Ritzer 2007, 85). In SST the central concepts of ST (agents, structures, duality of structure) have not only been molded in order to take into account contributions of theorists like Pierre Bourdieu, John B. Thompson, Margaret Archer, Mouzelis, and William Sewell Jr., but have been brought closer together in such a way that methodological explorations of specific social events are now made feasible. SST has been used in numerous studies ranging from organizational management (Jack and Kholeif 2007) and enterprise resource planning (Jack and Kholeif 2008) to childhood obesity (Chan, Deave and Greenhalgh 2010) and accounting (Coad and Herbert 2009).

In a nutshell, Stones argues that the duality of structure has four interrelated aspects: (a) the conditions of action (external structures, including rules, resources and position–practice relations); (b) embodied knowledge of actors (embodied "knowledge" and capabilities), which include conjuncturally specific knowledge related to the specific situation, and general dispositions acquired through previous socialization (what Bourdieu calls habitus); (c) active agency and agent's practices; and (d) the outcomes of action. The quadripartite cycle of structuration can be seen in Figure 4.2. I now turn to explaining in greater detail the theory's premises.

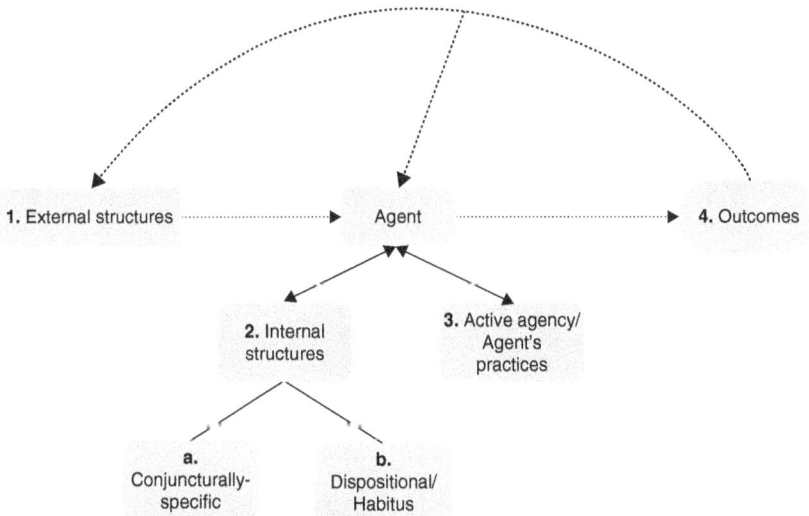

Figure 4.2 The quadripartite cycle of structuration.

External structures

External structures can be conceived as conditions of action and constitute
the structural context of action faced by the situated agent (agent-in-focus) at
the beginning of the specific social action (Time 1) (Stones 2005, 84). These
structures exert a causal force or influence on the agent and involve *position-
practices and their networked relations* along with *rules and resources* (material and
virtual), which at the same time constrain and enable the agent-in-focus to
"make history" by making use of the possibilities and capacities presented
(Stones 2005, 109).

External structures as constraints/enablements

When these external structures[6]—whose existence is autonomous from the
agent-in-focus—constitute the conditions of action, the agent's-in-focus
ability to act is affected in three major ways. First, the action is contingent on
the agent's ability or inability to enlist the aid of "objectively existing" structures of *domination/power* in order to pursue desired goals, or by being subject to the structural constraints targeted to noncompliant actors. Second,
by successfully deploying *normative* sanctions or rewards that are embedded
in the structures of legitimization or, conversely, by being subject to such
normative constraints. Third, by being enabled or constrained by the *interpretative schemes* of other agents that act within the same structural context
(Stones 2005, 60–61).

    In order to clarify what this means, one can take the example of a biologist
working in a laboratory of an agbiotech corporation who wishes to negotiate
for better working conditions (these may include issues of safety, a more flexible work schedule, upgraded equipment and so on). The biologist's decision
to voice concern is contingent on: (a) the existing structure of domination/
power; (b) the normative rewards or sanctions that such a move will trigger;
and (c) how other agents are going to interpret such an action. Therefore,
the external structures that confront the biologist are related to (a) how the
administration treats such requests (whether they are tolerant and open to
discussion or, for example, are more likely to follow an authoritarian trajectory by firing employees)—the presence/absence of an association or union
that supports such requests and, therefore, enables the biologist's demands
or not; (b) whether such a request is valid and has been realized before—if
there is an acceptable patterned way of carrying out this action; if enlisting
to the help of an association is the norm, and so forth; (c) How is this action
going to be interpreted by the administration? Is the biologist going to be construed as a threat? Is she going to be ignored? How are her fellow researchers

going to interpret this move? All these concerns constitute some of the external structures that face the biologist at Time 1 (before she acts). Depending on the situation they may comprise acknowledged conditions of action (i.e., the agent-in-focus thinks he is aware of them) or unacknowledged conditions of action (i.e., the agent-in-focus cannot give an answer to these questions, or there are other conditions that the actor is totally unaware of).

While Stones accepts the fact that for some agents certain external structures are intractable, at the same time there are cases in which seemingly "irresistible causal forces" can be malleable. In order for an agent to be able to resist the pressures of certain external forces, the following three types of properties have to be present:

- Adequate power or capability to resist the external pressures without jeopardizing the realization of core commitments that the agent-in-focus has to fulfill;
- Adequate knowledge of relevant external structures and of alternative routes of action and their respective ramifications;
- Adequate reflective distance from the conditions of action in order to take up a strategic stance in relation to particular external structure. (2005, 114–15)

All things being equal, one can notice that the greater the number of these qualities, the higher the likelihood that the agent will be able to regulate, modify or even erase specific aspects of external forces. The possession of such abilities is contingent on an agent's position within an organization hierarchy (e.g., whether a manager or a rank-and-file clerk), the agent's cultural capital, gender, race and so on (2005, 115). It is, therefore, very likely that the biologist of the previous example has far greater chances of succeeding if: (a) she has the adequate power to resist a given structural status quo as long as her core commitments (i.e., not risking being excluded from the social game) are not at stake; (b) she has ample knowledge of the objective environment she is confronted with and of possible trajectories of action along with their consequences; (c) there is a relative, but significant, ability for her to distance herself from the conditions of action for strategic and reflective monitoring.

Position-practices as external structures

Until now we have demonstrated that, according to SST, at the very beginning of a social action (Time 1), the agent-in-focus is situated in a social environment in which external structures enable or constrain her trajectories of action in certain ways. One facet of this external environment of action that influences, to a greater or smaller extent, the agent's action is comprised of position–practice relations. The properties of these position-practice relations entail:

(1) Positional identities that are clearly demarcated by identifying criteria such as documented qualifications and evident attributes;
(2) Clusters of practices through which identifying criteria, privileges and duties become apparent and acknowledged by other social actors;
(3) A variety of different position-practices that should be, or may be, inter-related with a particular position-practice;
(4) A range of mutual dependences, including asymmetrical power relations, through which position–practice relations unfold (Stones 2005, 62).

Two things regarding the nature of position-practices have to be clarified. First, while these four aspects of position-practices clearly have a dynamic character since, apart from the positional identities, prerogatives and obligations, and so forth, there is a vivid sense of *emergent/interactional* qualities. Nonetheless, at Time 1, the emergent properties of the position-practices can be seen as outcomes of previous interactions and, therefore, can be construed as preexistent conditions for subsequent actions. As a result, Stones argues, these positions and their respective practices once established—as previous actions have unfolded over time and space—will preexist the particular agents of action that subsequently occupy them. Second, and following from the previous point, it should be clear that position-practices can be identified regardless of *who* occupies a specific social position at a given time. A position-practice (for example, that of a chief scientific adviser, that of a geneticist, that of a farmer and so on) carries with it a set of identifying criteria, duties and obligations, documented qualifications, mutual dependences with other position-practices (CSA and MPs, geneticist and other lab assistants, farmer and seed provider respectively) and so forth irrespective of its incumbent. Albeit, it must be remembered that the production, reproduction and transformation of these position-practices is "contingent, *inter alia*, on the activity of position-taking and making and is by no means automatic" (Stones 2005, 63, stress in original).

Resources as external structures

It is not only position-practices that form the external environment of action, but also the presence and availability of resources. This is the second and final major component of what constitutes the external structural environment at Time 1. For Stones, resources can be construed as both virtual and material. Resources as external structures, together with position–practice relations, constitute the objective environment of action that the situated agent faces at the beginning of a social action (Time 1). These resources enable and/or constrain the agent who can or cannot draw upon them as the action unfolds over time and space. They have both a material and a phenomenological connotation, as they refer to materials and/or ideas. They can be plain raw material, tools, technologically sophisticated machines or can include the existing structures of power, meaning and norms (see Stones 2005, 35–36). If we take the mundane example of a farmer growing GM crops, his actions at the time of sowing will be heavily contingent on the existing structural environment at Time 1. The material resources the farmer has to take into account—which can or cannot be controlled—could be the current quality of the soil, the traits of the seed, the available sowing machine and seed drills, the weather conditions and so on. The virtual resources that also constitute the environment of action that affects the farmer's actions— and which he may or may not utilize in one way or another—can include market trends and the demand for specific products, the royalties paid to the company that has provided the seeds and the specific procedure that has to be followed (power), the stance that the farmer's family and immediate environ-ment have toward GMOs (meaning), the typical farming practices (norms) of fellow farmers and so on. Therefore, at Time 1 the situated agent is faced with an objective structural environment that consists of resources (mate-rial and virtual) the agent may or may not use as the action starts to unfold. Having examined the role that external structures play in providing the envi-ronment of action, I now turn to examine what internal structures are and how they affect social action from their part.

Internal structures

The internal structures of SST are drawn upon as the medium of the agent's conduct and consist of the conjuncturally specific knowledge of external structures and the general-dispositional characteristics of the social actor or his/her habitus (Stones 2005, 86). Both are virtual, and both have a strong phenomenological connotation, as they are responsible for how the external structures are interpreted, construed and, eventually, acted upon. Moving on to internal structures is of crucial importance to SST, as in this way we move from a *potential* ontology-in-general that external structures have to an ontology-in-situ, where specific structures and agents in specific locations and time instances are examined. At this level, one is in some sense obliged to ask the questions: "who," "what," "where," "when," "why" and "how."

The conjuncturally specific knowledge of external structures

The knowledge the actor possesses of the external environment that exists at Time 1 is of great importance for the ways that a social action will unfold over time and space. The conjuncturally specific internal structures entail "stocks of knowledge" directed out toward the context of action. They are neither transposable nor generalized schemata. "They have contours, shapes and textures whose *specificity* within time and place is of great import to the agent facing those external structures" (Stones 2005, 90, stress added). This kind of internal structure refers to the ways an agent perceives a *specific* context action at a *particular* time and place. It refers to the ways a Rothamsted Research intern, for example, knows how to dress, sit, talk or ask questions in a lunchtime seminar, and to her knowledge that while this type of behavior is endorsed in that specific setting, it is rather unacceptable during the presentation of her findings to a BBSRC panel. While conjuncturally specific knowledge refers to specific contexts, it does not follow that this type of knowledge was gained at that point and place in time. It may have been formed well in advance, and it may be relatively enduring knowledge. Alternatively, it may have been founded previously but refined during the course of interaction.

Since this type of internal structure forms one window that the situated agent has toward the outside world of structures, conjuncturally specific knowledge can be divided into knowledge of: (a) interpretative schemes (i.e., what will other people think and how they will react); (b) power capacities (power resources available and how the actors and other agents are going to utilize their access); and (c) normative expectations and principles of the agents within context (i.e., what is an acceptable/plausible patterned way of behavior that is going to be followed) (Stones 2005, 91–92).

The general-dispositional/habitus

Conjuncturally specific knowledge is one facet of the internal structures that an agent has. The other is the general-dispositional, or what Bourdieu has coined as *habitus*. For Bourdieu, the habitus refers to "the durably installed generative principle of regulated improvisations[, which produces] practices" (1977, 78). The habitus refers to the partly unconscious and "can be understood as the values and dispositions gained from our cultural history that generally stay with us across contexts (*they are durable and transposable*) (Webb et al. 2002, 36, stress added).

> The habitus, a product of history, produces individual and collective practices—more history—in accordance with the schemes generated by

history. It ensures the active presence of past experiences, which, depos-
ited in each organism in the form of schemes of perception, thought and
action, tend to guarantee the "correctness" of practices and their con-
stancy over time, more reliably than all formal rules and explicit norms.
(Bourdieu 1990, 54)

These generative schemata of action, which are durable and transposable products of prior socialization, include typifications of things, habits
of speech and gesture, typified recipes of action, classifications, generalized
worldviews and methodologies for adapting this generalized knowledge to a
range of particular practices in specific settings in time and space. The habitus is drawn on naturally, without thinking, in most aspects of an agent's social
life (Stones 2005, 88). What this tells us is that one's outlook on life (whether
one is ambitious or not, whether one sees society as full of opportunities or
merely inequalities), one's views on technological progress, homosexuality,
Communism, one's table manners, the clothes one wears and so on, are heavily influenced by the habitus: the deeply embedded generative schemata of
action/dispositions that have been acquired through prior socialization. In
other words, all of the above attitudes and behaviors are severely affected by
one's social class, race, gender, religion, the ways in which these structures
have been internalized, the family one has been brought up to, education
received, significant others who have influenced the agent's-in-focus person-
ality and so forth.

Active agency and agent's practices

Apart from the presence of internal structures, it is the agent's practices that
complete the puzzle of how an agent acts in context, as it is during a social
interaction that the agent's internal structures combine—either consciously
for strategic reasons or unconsciously—are molded and take final shape
(Stones 2005, 100–1). Five key aspects are important for the analysis of an
agent's practices. Namely:

- The shifting *horizon of action* plays a role in the ways that agents-in-focus is
  going to draw on their internal structures and explore trajectories that—
  under different circumstances—may not have been explored;
- Aspects of creativity, innovation and improvisation: agents' conduct
  or practices are also influenced by their likelihood to be creative and
  innovative. This does not refer to contextless or ahistorical instances of
  improvisation, but rather to instances in which the skills, habits and orientations of one's habitus are combined with the demands of the current

conjuncture and unfold in open and indeterminate ways (Stones 2005, 101–2).

- The degree of critical distance: that is, for the agent to be aware of the horizon of action, there certainly has to be a degree of critical distance between agents and their own internal structures (Stones 2005, 102). It can be the case that agents will not draw on their internal structures in a taken-for-granted manner but may distance themselves from the virtual order of structures for strategic-monitoring reasons;
- The hierarchization of concerns and priorities means that the agents' conduct is rationalized by their sorting of concerns, priorities and purposes;
- The conscious and unconscious motivations play a "relatively autonomous and dynamic role in agent's conduct" (Structuration Theory 2005, 102).

Outcomes

The final part of the structuration cycle is naturally the outcomes of the action. These outcomes need to be construed not only as events but also as effects on internal and external structures. These effects can have a negligible impact on structures; they may lead to their reproduction and consolidation, to their elaboration or to their change. In the previous analysis of internal structures, the discussion evolved around their use as a medium, as part of the agent-structure duality that agents draw upon during their conduct. When speaking about outcomes, however, the internal and external structures constitute the result of the action and, therefore, constitute the new external and internal structural environments of action that will exist at Time 1, when the new cycle of structuration starts to unfold.

Conclusion

In this chapter, some of the key contributions of Benton, Mouzelis and Stones, pertinent to the research question at hand, were discussed. While these theorists argue from their own perspective and overarching theoretical positioning, throughout their works they have all encouraged caution to deterministic and reductionist accounts often found in the discipline at large and the subfield of STS in particular. All three have also offered their own distinct, but theoretically and methodologically compatible, exegeses as to how particular impasses can be overcome. To reiterate, it was argued that from Benton's rich and multi-disciplinary work, the concepts of ontological naturalism and the hierarchization of sciences can sensitize the researcher to a more balanced approach regarding the human/nonhuman relations in

STS: a problematic often conceptualized in either deterministic or confla-tionary terms. Mouzelis's post-Marxist scheme, at the same time, manages to break away from the economic determinism that is usually part and parcel of Marxist theorizing and offer a holistic approach to the study of institu-tional spheres. Finally, Stones has refined Giddens's Structuration Theory from the level of ontology-in-general down to the ontic in a way that enables the researcher to examine specific interactions as they unfold over time and space within explicit contexts. What remains to be seen now is how these three approaches can be brought closer together in an ontologically sound and methodologically coherent manner.

Chapter 5

A HOLISTIC FRAMEWORK FOR THE STUDY OF AGRICULTURAL BIOTECHNOLOGY

In Chapter 4, the contributions of Benton, Mouzelis and Stones, relevant to the broader research question at hand, were presented to the reader. In this chapter, these contributions will be reappraised, modified whenever necessary, and realigned so as to function as interrelated parts of a broader holistic framework the aim of which is to examine the sphere of agricultural biotechnology from an externalist and internalist perspective. The chapter starts with a reassessment and modification of Mouzelis's TAI scheme and then moves on to revisit another version of Stones's SST, proposed by the British sociologist himself, which incorporates the technological element. Within this version of SST, Benton's ontological naturalism is integrated, and this new scheme is meshed with the modified version of TAI. At all times, this critical synthesis is informed by clear ontological premises and specific methodological brackets.

TAI and the Field of Agribiotechnology: Criticisms and Modifications

While the TAI scheme appears to offer interesting insights when it comes to examining changes in major institutional spheres in the *longue durée*, still there are some modifications that have to be made in order for it to be applicable to cases in which the unit of analysis is not a major institutional sphere (i.e., economy, polity or culture). By examining the field of agbiotech, and more precisely the case of GM seeds or food, the level of analysis becomes both narrower and broader. It becomes narrower in the sense that agribiotechnology is an institutional sphere that does not stretch as wide in time and space as the economy, polity and culture and, therefore, does not play such a fundamental role so as to influence, to a greater or lesser extent, the everyday lives of individuals. At the same time the scope of analysis becomes wider because, despite the relative autonomy of (techno)science, as Merton (1942) rightly argued, it is a field in which social forces from the major institutional spheres intersect and form complex figurational and institutional ensembles. This is so because, although this sphere has its own dynamics and internal logic, the fact that it is new renders it more prone to external influences. In other words, it is impossible to examine agribiotechnology in its own right by bracketing the social pressures stemming from other fields. The purpose of this section is twofold. First, I will suggest one way of modifying TAI so as to make it more appropriate for the assessment of science and technology. Second, while doing so, I will also raise and address some issues of concern found in the original framework.

Technology in agribiotechnology

As we saw in the previous chapter, Mouzelis, following Cohen (1978), uses Technology in a broad sense that is not restricted to technological artifacts. Rather, Technology or productive forces refer to instruments of production (tools, machines, premises and instrumental materials), labor power, knowledge and technical organizational arrangements (Cohen 1978, 55). The presence of humans and any form of socialness are *by definition* excluded in this offering of forces of production. What is also very significant, but equally troublesome, is that these forces or powers are objects and do not entail any sort of relation (Cohen 1978, 28). We are presented, therefore, with a definition—which Cohen proposes and Mouzelis gladly accepts—where forces of production refer to entities that remain abstract and unrelated to one another. I find it tough to feel comfortable with such a definition, as the presence of human actors and social interactions are *clearly* present in "labour power, knowledge, and technical organizational arrangements"; after all, we speak of *A*'s ability to work in a specific way, the knowledge that *this* person has, the technical arrangements among *those* workers decided by *that* manager and so forth. By forcefully deciding to deprive productive forces of the presence of humans leads us to a lopsided view of social reality by espousing a form of system-essentialism—something Mouzelis has so much wanted to avoid. Cohen's adamant rejection of any form of social relations in the productive forces makes it impossible to theorize any sort of interaction within Technology.

While Cohen's interpretation of Marx leads him to strip productive forces of any social mantle, Derek Sayer's (1987) understanding is radically different. While debates on even the most fundamental Marxist notions can be very extensive and complicated, at this point I will simply limit the discussion to the points raised by Sayer, which assert the social dimension of productive forces. Sayer accuses Cohen of grossly misinterpreting the nature of productive forces and criticizes him for adopting a viewpoint that eventually leads to a "mystification" or "fetishism" of the productive forces (Sayer 1987, 20–21). In fact, one of the main recurring points in Marx's writings is his insistence that production has an irreducible social nature. Certain passages in Marx's works, Sayer argues, are simply undeniable.

The production of life, both of one's own in labour and of fresh life in procreation, now appears as a twofold relation: on the one hand as a natural, on the other as a social relation—social in the sense that it denotes the co-operation of several individuals, no matter under what conditions, in what manner and to what end. It follows from this that a certain mode of production, or industrial stage, is always combined with a certain mode of co-operation, or social stage, and this mode of co-operation is itself a "productive force." (Marx 1998, 49)

If, therefore, Sayer's reading of Marx points toward an inherent social dimension of productive forces, then the initial impasse of including social interactions in Technology is avoided. By construing forces of production not merely as a list of abstract, inanimate tools used during the production process (i.e., productive powers of the *material conditions of labor*), but as the productive powers of *social labor* (Sayer 1987, 26), then the possibility of discerning social relations becomes plausible.

Having argued for the existence of *social relations* in Technology, we can start modifying TAI for the field of agbiotech. In the initial formulation of the framework, Mouzelis referred to technologies or forces of production that can be discerned in each major institutional sphere (economy, polity, culture/religion), and which shape, to a smaller or greater extent, the lives of virtually all people within a state. Agbiotech, on the other hand, is a minor field, and quite obviously such equivalently broad and all-encompassing normative and institutional arrangements cannot be found. It seems to me that one way of tackling this issue is by arguing that Mouzelis is theorizing at a level of abstract ontology focusing on broad spheres of society without investigating the details of their relations with each other, which one would have to do in order to inquire into the unfolding of processes within concrete conjunctures. Because concrete, situated processes involve the articulation of elements from different spheres, our case can benefit from a more situated analysis. This requires the introduction of concepts that can mediate between the abstract and the concrete, *in situ*, level. If this suggestion makes sense, then what follows is that we need to start placing Technology of agribiotechnology in more situated contexts. Since in our case Technology refers to the social forces and relations that produce GM seeds, we may turn our attention to the social locations in which these GM seeds are produced. There appear to be two main loci where the production of GM seeds takes place:

- Places of Research and Development (R&D).
- Places of cultivation (farms).

If this suggestion is coupled with the previous one (i.e., that productive forces have a social and relational character), then we can start talking about a variety of relations that unfold within these social locations. Taking into account that apart from humans, productive forces include machines, apparatuses, tools, genes, chemical compounds and so forth we can speak of two kinds of relations: social and sociotechnical. The former refer to human/human relations and the latter to human/nonhuman relations. I will try to develop these two conceptions of relations a bit further in the remainder of this subsection.

R&D can refer to places located at an independent laboratory, a subdivision of a transnational corporation, a university campus, a publicly funded research center and so on. In such locations, GM seeds are designed and tested in some form or another; that is, R&D may test a particular genetic sequence introduced into living cells using microprojectile bombardment; it may test for unintentional mutation of cells after tissue culture; it may test some traits of the GM seed/plant itself in a closed environment and so forth. During the process of research and development, scientists engage in *sociotechnical relations*. That is, they interact with nonhuman entities in their aim to achieve a specific goal. Some of these nonhumans may be tools with well-known functions and operations, such as temperature-, pressure- or acidity-measuring machines that are used on a routine basis, but other nonhumans may be living organisms whose behavior and responses to external changes or stimuli may be unpredictable. In R&D we can also discern *social relations*; that is, relations among scientists regarding the research and development of GM seeds. Such relations can be both orderly and conflictual. They can be relations of formal and informal cooperation by taking the form of joint ventures, of friendly information and advice exchange, of parallel project development and so on. They can also be (symbolically) conflictual by revealing disagreements or tensions regarding the specificities of GM seeds in R&D. For instance, scientists or scientific groups may disagree about the nutritional value of a GM plant, about its ability to irreversibly transfer its genes to non-GM plants in the open environment and thereby creating new hybrids, about the lifespan of the "volunteer" GM seeds in the soil and much more.

In *farming*, too, one can discern sociotechnical and social relations. *Sociotechnical* relations refer to the ways in which the farmer interacts with the seed/plant. Such relations can refer to the amount of fertilizer farmers use, to the ways in which the plants respond to such usage, to the effects that weather conditions have on the plants, to how farmers decide to accommodate such reactions from plants and so on. *Social relations* in farming refer to the relations farmers develop among themselves, which can be related to, among other things, the cultivation/production of GM seeds/plants/crops. Such relations may revolve around issues of problem analysis, suggested techniques based on prior general knowledge or current trial and error analysis, opinion exchange on cultivation and more.

Appropriation in agribiotechnology

While in the initial formulation Appropriation and Technology appear to be interrelated only to the extent that the former represents the social actors (collective and individual) who compete to control the latter, I would like to argue that the relationship between the two is not as haphazard as it may appear to be. The mere presence of social actors attempting to appropriate Technology signifies that, for one reason or another, these actors have vested interests in controlling these forces of production. What also holds true is that these groups obviously want to increase their chances of successful appropriation. Now, I do not see any apparent reason we should ignore the possibility that *sometimes* these vested interests are crystallized as structural arrangements present in Technology. These structural arrangements constitute, to some extent, the external environment of action for the productive forces and shape the way that interactions unfold so as to increase the chances of the desired outcome and the subsequent control. Therefore, very often power relations are found to be inscribed within the social and sociotechnical relations of Technology. When this is the case, the additional epistemological focus of Appropriation helps bring to the surface such issues of power that influence these relations. On other occasions, of course, Appropriation may not be so influential and penetrative. In such cases, social actors may limit their attempt to control technological artifacts without exercising any influence on the conditions of action. I will devote this remaining subsection by clarifying the argument that has just been suggested.

If, for our field of interest, Technology can be located in R&D and farming, then it logically follows that social groups that try to control Technology are actually or virtually gathered at least at these two loci. Appropriation in *R&D* can be structurally penetrative and set the conditions of action in these social locations. Such conditions may be manifold: they may be related to the allocation of funds that social actors decide should be made available to a particular R&D scientific team; they may have to do with the decision to facilitate the scientists' work by upgrading the equipment used; with the agenda-setting of a research center and more. All such examples show how, on certain occasions, decisions to appropriate Technology shape the environment of action of scientists and other social actors working in R&D by making funds available to them, with the presence of better equipment, a specific research topic and a time deadline to be met and so on. All these are resources and structural arrangements which, in many cases, scientists take for granted. What has also become apparent is that social groups that seek to appropriate Technology are not necessarily actors in the same field. The above decisions can be made not only by agribiotech firms, but also by charity organizations, research councils, university departments and much more. Therefore, actions stemming from different institutional spheres may function as conditions of enablement or constraint for the productive forces. Appropriation in R&D can, of course, be subtler, as in the case of patenting, where attempts to control Technology do not take such prominent dimensions. Also, it is worth mentioning that Appropriation can take non-institutionalized forms in instances of protest where GM seeds and crops, cultivated either for experimental reasons or commercial use, are destroyed as a symbolic gesture by the protesters asserting that this technology should be banned.

Moving on to Appropriation in *farming*, there too we can discern actions of social groups and individuals who try to control the agricultural process and set structural conditions to the farmers' actions. If one takes into account, for example, how farmers who cultivate GM crops are in essence obliged to use specific "companion" chemicals, and how GM seeds constitute patented corporate property that cannot be traded between farmers, either intentionally or unintentionally,[1] it becomes very obvious that farmers act within a highly regulated environment (Kloppenburg 2010).

Apart from the two social locations mentioned, it seems there is a third social space in which Appropriation takes place. Since GM seeds are commodities that, after leaving the laboratory and the farm, are sold, bought and potentially consumed (as GM food), one can also find Appropriative forces in the process of *distribution*. This constitutes a very rich and diverse social trajectory that encompasses many aspects, ranging from legislation about the cultivation and commercial use of GM crops and rules of traceability and labeling, to the opening up of markets previously unwilling to accept GMOs and the decision of supermarkets to shelve or ban GM foods. In such locations, one can notice (symbolic) altercations—such as the ones briefly discussed in Chapter 1—among agbiotech firms, NGOs, concerned individuals, trade unions, farmers' associations and much more, who try to acquire favorable legislation or conditions of action so as to advance their own goals.

Ideology in agribiotechnology

To repeat, for Mouzelis, ideology refers to the discursive and non-discursive practices that:

- give a distorted picture of how the means of production are owned/controlled;
- provide reasons to justify and legitimize the prevailing ownership/control arrangements (1990, 68).

The first suggested function of ideological practices rests on two fundamental concepts; the idea of *distortion* and the presence of *means of production*. I think that the presence of both these concepts is a source of concern.

First, Mouzelis's acceptance that ideology offers "a distorted picture" of the relations of production strikes as a direct reference to Marx's *camera obscura* statement; that is, "in all ideology men and their circumstances appear upside down as in a *camera obscura*" (Marx 1998, 42). Roughly speaking, one can conceive a central lineage from Hegel and Marx to Georg Lukàcs and some later Marxist thinkers who have been preoccupied with the ideas of illusion, distortion, mystification and true and false cognition (Eagleton 1991, 3). While these notions of distortion, false beliefs, illusion and so on have been extremely influential, it is very difficult to overlook some major weaknesses they carry. For one thing, if ideology is said to offer a distorted, obscure or simply false picture of "the way things are," then it logically follows that there is some unmistakably right way of seeing things. But, if this holds true, who manages to see things correctly and on what grounds? I think Terry Eagleton, in essence, is right in arguing that "the belief that a minority of theorists monopolize a scientifically grounded knowledge of how society is, while the rest of us blunder around in some fog of false consciousness, does not particularly endear itself to the democratic sensibility" (1991, 10–11). This idea of distortion and falsehood in ideology was certainly not upheld in either Gramsci or the structural Marxism of Althusser (McLellan 1995, 28). For Althusser, ideology is not related to any illusory representations of reality but "is a matter of the *lived* relation between men and their world" (1969, 233; stress in the original).

Second, it is very difficult for me to find convincing reasons why ideology has to be necessarily related to the means of production. In fact, Althusser clearly expressed the need to detach ideology from means of production by suggesting that "ideology arises before class divisions appear, and will survive after these divisions disappear. Ideology is a structural feature of any society" (Larrain 1979, 156). For the French philosopher, "*ideology is eternal,* exactly like the unconscious" (Althusser 1971, 161; stress in the original). In various xenophobic, sexist or homophobic utterances, are arrangements of the means of (cultural) production entailed or is someone expressing a much shallower view of the world? If someone believes "Crying is a sign of weakness," to what productive means are they referring? Is the utterance, "We have suffered a lot because of the French," a necessarily deeply political stance that has to do with the means of political/military production or could it be a discursive element whose significance will remain only in the field of sentiments and emotions?

Not only point (a) in Mouzelis's suggestion appears to be problematic, but also point (b) seems to be somewhat inaccurate. Even if the persistent reference to productive means and their arrangement is bracketed, we are presented with a proposition that portrays ideological constructs as a privilege, tool or even a cunning technique used *exclusively* by the ruling or powerful groups that have vested interests in maintaining the current arrangements. It is rather obvious that such a suggestion seriously undermines the possibility that the "ruled," less-powerful groups employ their own ideological constructs that construe facets of the social world in a different manner (Abercrombie and Turner 1978). By arguing that discursive elements that may attribute all sorts of political, cultural and aesthetic malaise to "plutocracy" and the "needs of capital" do not constitute ideological propositions, but more or less "objective" or "scientific" judgments, one is basically retracting to the traditional Marxist matrix consisting of ruling/ruled, true/false, right/wrong and so forth. I hope it has been convincingly portrayed by now that ideology needs to be discussed in ways that extend much further from such unsparingly dichotomous reasoning.

   If, therefore, the idea of distortion is dropped and the obligatory reference to the means of production is eschewed, what is left of ideology? Despite the controversy surrounding ideology, it seems to me that some headway can be made by spelling out some of the term's key constitutive components, which appear to be largely uncontentious. It appears that a minimal common denominator can be found in various theories of ideology with the following characteristics:

- Ideology has a *relational* character in the sense that in the ideological there is always a confrontation between "subject" and "object" (Eagleton 1979, 62; Warren 1981). Also, "ideologies are shared by groups or populations that have collectively and individually shared experiences" (Fine and Sandstrom 1993, 26). These shared experiences—but also shared choices, morality and politics—have a *collective implication* about some sort of shared commitment (Craib 1981, 508–9).
- Ideological propositions are *apparently* propositions about the real. They are pseudo-, para- or virtual statements about a "real" entity, issue, problem, arrangement or state of affairs—expressed in a way that is not necessarily wrong (Eagleton 1979, 64). These statements may be blatantly erroneous, of course, but may also be partially true, or true under very particular circumstances or conditions that are not necessarily mentioned. In other words, ideological discourse may be true at one level, but not at another. It may be "true in its empirical content but deceptive in its force, or true in its surface meaning but false in its underlying assumptions" (Eagleton 1991, 16–17).
- Ideological constructs can be decoded into *emotive* and/or *action-oriented* discourse (Gouldner 1976; Habermas 1976; Fine and Sandstrom 1993). Seen in this light, ideological propositions differ from referential enunciations in that the latter do not allude to the subject's emotions or attitudes (Eagleton 1979, 64).

- Ideology is an aspect of "every system of signs and symbols in so far as they are implicated in an asymmetrical distribution of power and resources" (McLellan 1995, 83). What makes discourses ideological is their connection with systems of power/domination (Purvis and Hunt 1993, 497).
- By combining points (c) and (d), we can argue that *sometimes* ideology is used in order to trigger some emotional or action-oriented response related to the asymmetrical distribution of power and resources. Such responses may help sustain, facilitate or challenge the existing or emerging distribution of power and resources.

If these five propositions are accepted as key elements of ideology, then we can start bringing them closer together so as to clarify what can qualify as an ideological construct in agribiotechnology. A tentative definition may be the following: *Ideology in agribiotechnology refers to the discursive elements shared by members of a group or population that may trigger sentiments or actions of human agents, who may not necessarily belong to that group/population, aimed at supporting or opposing the research, production and/or distribution of GMOs in its current form.* With this definition, I have tried to assimilate largely uncontested key ideological components into a coherent and ontologically situated articulation. At the same time, I have also attempted to retain ideology's critical edge by making obvious reference to the fact that the "current" (no matter how enduring this may be) arrangements in research, production and/or distribution of GMOs can be the source of (symbolic) conflict. The fact that there may be competing ideologies (i.e., broadly speaking pro- and anti-GMO) does not necessarily mean that social actors can be divided into corresponding groups. There may be actors who support GMOs but do not espouse all ideologies in favor of this technology and vice versa. Also, there may be skeptics who are not convinced of the overall contribution of GMOs, but nonetheless do not accept many of the anti-GMO ideological propositions.

Having said this, we can move on to mention briefly a few examples of ideological constructs used in agribiotechnology. As previously mentioned, such discursive elements may be quite narrow and concrete in the sense that, at first glance, they *appear* to be objective referential statements and/or technical terms. However, with closer analysis, they prove to have little to do with objectivity and more to do with serving specific goals and reflecting specific frames of mind. One such example is the notion of *substantial equivalence*, which suggests that for all practical purposes GMOs can be considered to be equivalent to their "conventional" non-GM counterparts. While substantial equivalence is referred to as a principle, in fact, it is a highly debatable discursive construct that is interpreted quite differently in the EU and the United States (Welsh and Ervin 2006). North American regulatory regimes espouse this principle while the EU maintains that there should be a fundamental distinction in regulation regarding GMOs and natural organisms (Levidow and Murphy 2006, 119–46; Lezaun 2006).

Ideology can, of course, be discerned in much broader utterances about the (social) environment. Some key discursive elements articulated in the GMO debate are those of democracy, elimination of world hunger, fight against poverty, progress, preservation of nature and more. These (grand) narratives are, quite naturally, highly contested in the GM debate (e.g., EuropaBio 2009, 2010; Friends of the Earth 2007, 2010; Kaye-Blake et al., 2008) not in relation to their surface value, of course, but with respect to their deeper significations and connotations.

Conclusion: Key Points in the Adapted TAI Scheme

In this section, I have suggested one way in which the TAI scheme can be adapted to the field of agbiotech. Apart from arguing for a more situated ontology that will help assess a minor institutional field, I have also tried to address some issues that appear to be somewhat problematic in the initial TAI. Regarding Technology, it has been argued that the productive forces have an undeniable social and relational character. Next, I have tried to demonstrate that there is some link between Technology and Appropriation that needs to be highlighted in the sense that, quite often during Appropriation, social actors shape the external environment of the productive forces. Lastly, in Ideology, I have argued against the typical Marxist connotations of concealment and falsehood on the grounds that it is too limited a definition, and I argued for a broader, but still critical, concept of ideological discourses.

What has been mentioned a few times during the analysis of TAI is that, despite the attempt to make this post-Marxist framework more specific, there is a pressing need for a complementary theory of action that will help shed light on greater details of how social and sociotechnical interactions unfold in agribiotechnology. I now turn to address precisely this issue.

Strong Structuration Theory and Technology

Five years after the publication of *Structuration Theory*, Stones (Stones and Greenhalgh 2010) provided a theoretical explanation regarding the effects the UK's National Health Service IT programs have on the macro, meso and micro levels. In order to do so, parallels were drawn between SST and ANT, and the technological dimension was added to SST (Stones and Greenhalgh 2010, 1285). The structuration cycle with the technological aspect incorporated reads as in Figure 5.1.

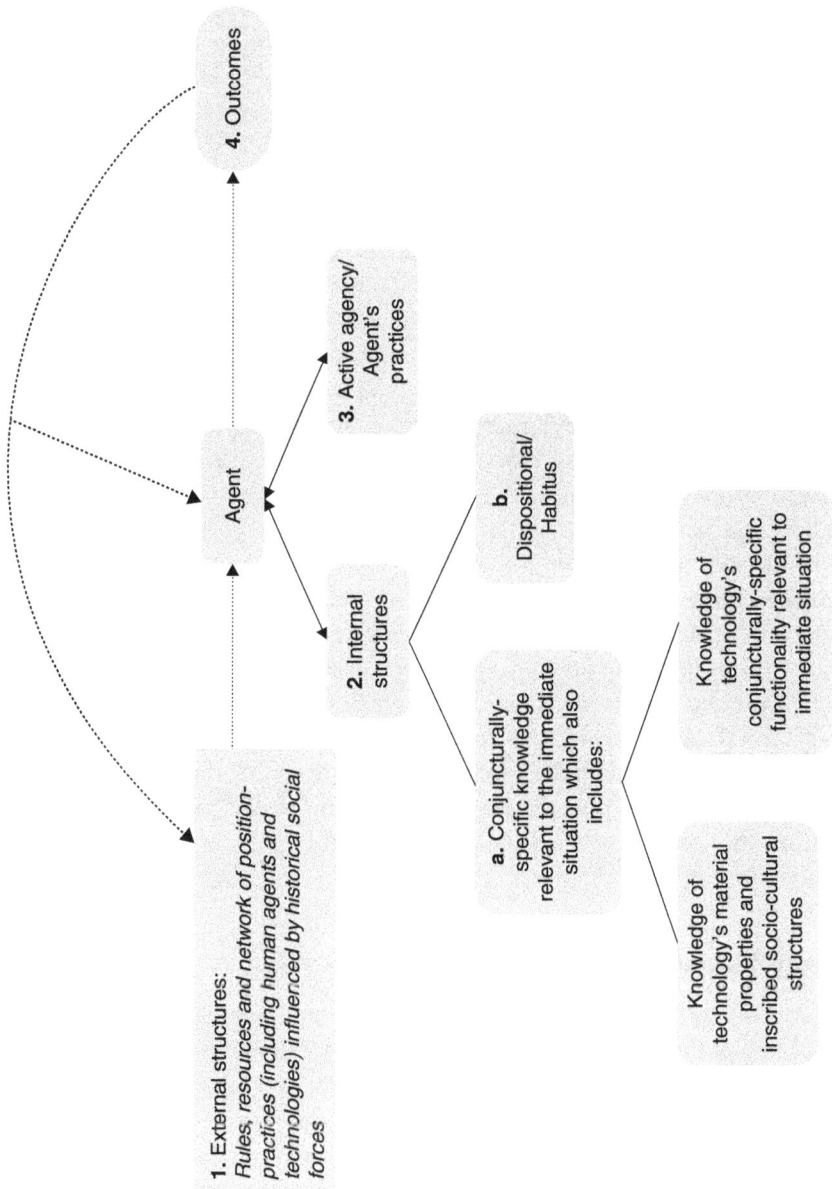

Figure 5.1 The quadripartite cycle of structuration with the dimension of technology.

What is, in essence, added in SST[2] is the acceptance that: (a) technologies also play a part in the constitution of the environment of action; (b) these technologies have material properties and inscribed sociocultural structures; (c) they have a conjuncturally specific functionality relevant to the immediate situation; (d) the agent-in-focus has more or less knowledge of points (a), (b) and (c). It is not difficult to notice what this means in practice. The first refers to the obvious realization that the environment of action not only consists of position–practice relations and rules and resources, but also of technological artifacts, such as the laboratory equipment a researcher has at her disposal or the machines and tools a farmer possesses in order to carry out tasks more efficiently. These technologies, of course, are neither dead material—that is, infinitely malleable, as social constructivism would have it—nor nonhuman agents, as Pickering would argue; rather, they possess material properties and inscribed sociocultural structures. This can refer, for example, to the materials used for the construction of two different cryogenic systems, and to the kind of versatility this difference in materials offers. However, it is not only the material properties, but also the inscribed sociocultural structures; the symbols that have been placed on machines, which have a specific meaning to the user, the color red on a button that terminates operations and so forth. Third, apart from these general characteristics, technologies also have conjuncturally specific functions relevant to specific situations; the specific drop-down menus that appear when a particular option is chosen in a precision agtech software program or the indicator that lights up when something is wrong with the hydraulic pump of a tractor and so forth. Finally, the situated agent is more or less aware of the technologies' ontological properties and the degree of her knowledge—which is different from that of the external observer—should be a matter of empirical investigation.

Placing Benton's ontological naturalism in SST

In the previous chapter, we discussed how Benton's ontological naturalism helps alleviate the human/nonhuman tension couched in most studies of science and technology. We also argued that his notion of embeddedness would be brought forward in this work. It is true that Mouzelis and Stones have both argued for the need to situate actors and interactions within specific economic, political and cultural contexts. Stones, as we previously saw, has gone to a further extent by placing technological artifacts in a situated ontology that recognizes the role their distinctive properties and functionalities play in the way interactions unfold. This is very telling about the human/technological artifact interaction, but since in agbiotech human agents interact not only with technologies but also with living organisms, some further elaboration may prove to be useful.

For Benton there are three main interconnected features that humans share with all living organisms: (a) they all need to interact with external nature in order to survive; (b) they all have their distinctive modes or patterns of interaction with nature, that is their "species-life"; (c) each species manifests its essential nature through its participation in its characteristic species-life (Benton 1993, 46). If this is taken as the founding block for the similarities of all living organisms, then there is no apparent reason we should not incorporate this idea in SST. In effect, the structuration cycle could take the following form (Figure 5.2).

So, in the examination of a scientist's interaction with a GM plant,[3] the researcher should take into account the scientist's knowledge of its distinctive mode of interaction with nature (among others, the distinctive characteristics it has been genetically modified to display), its generalizable and conjuncturally specific knowledge of the GM plant's needs for survival (quality of soil, resilience to specific weather conditions, amount and type of pesticide it can tolerate and so forth), and its general natural tendencies and the degree to which specific conditions constrain or facilitate these (probability to cross-pollinate with other types of plants, likelihood to yield specific amounts of crops, lifespan and more). It is worth repeating here that the degree of knowledge the scientist may possess may be high, but also very low, and the outcomes of the interaction may be desired and undesired, expected and unexpected.

It is hoped that it is becoming apparent that construing human/nonhuman relationships in this way can offer much greater insight into the ways the two interact. By combining at this stage the contribution of Benton and Stones, it seems we have avoided reductionist and conflationist techniques that so often appear in studies of science and technology. Humans are not attributed with any quasi-Promethean powers, but are placed in a structural environment in which they can act; technologies are theorized as having their own properties and functionality; and living organisms have been conceptualized as having their own needs, tendencies, and unique characteristics.

Until now we have placed humans, technologies and living organisms on a continuum and have demonstrated how they interact with one another. We can now turn to see how SST can be placed in the more general TAI scheme.

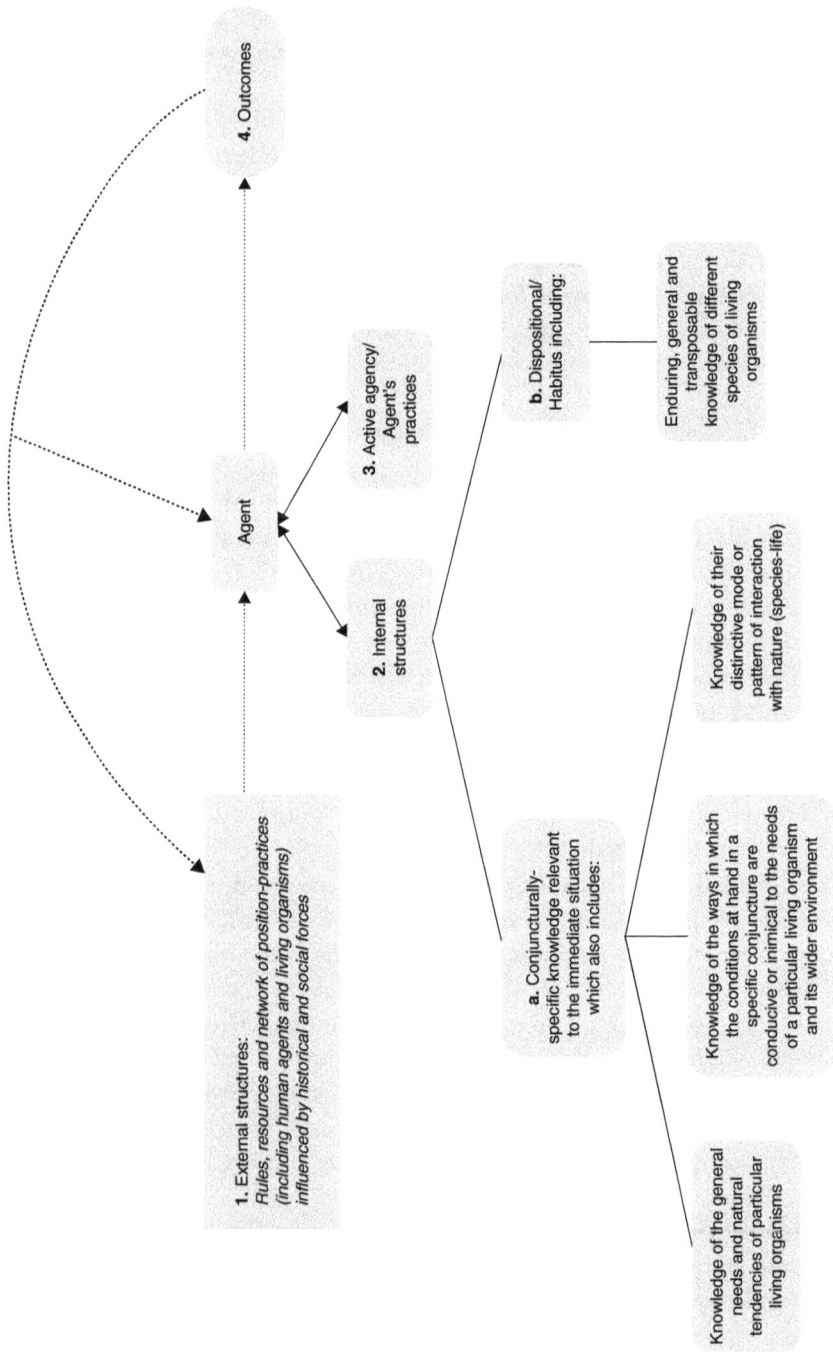

Figure 5.2 The quadripartite cycle of structuration with the dimension of living organisms.

4. Outcomes

Agent

3. Active agency/ Agent's practices

2. Internal structures

b. Dispositional/ Habitus including:

Enduring, general and transposable knowledge of different species of living organisms

1. External structures:
Rules, resources and network of position-practices (including human agents and living organisms) influenced by historical and social forces

a. Conjuncturally-specific knowledge relevant to the immediate situation which also includes:

Knowledge of their distinctive mode or pattern of interaction with nature (species-life)

Knowledge of the ways in which the conditions at hand in a specific conjuncture are conducive or inimical to the needs of a particular living organism and its wider environment

Knowledge of the general needs and natural tendencies of particular living organisms

Bringing TAI and SST Closer Together

A general remark that one can make regarding the Modified TAI and the SST and Technology/Living Organisms frameworks is that neither constitutes an end product, a theory of "something," or a fully worked theoretical construct that offers "knowledge" about the social world. Rather, they form a set of frameworks that help the researcher examine the social world from two different viewpoints. The post-Marxist framework argues that a fruitful discussion can be realized by dividing an institutional sphere into its three main dimensions while SST and Technology/Living Organisms focuses on greater details of social interactions by placing the notion of duality of structure at its core. Both, therefore, turn attention to different levels of the social world; the former is interested in broader time–space locations while the latter is more meso/micro oriented. If this holds true, then it becomes obvious that there is no apparent reason why one should not endeavor to combine the two so as to examine various levels of abstraction and differing interactions that unfold within a particular institutional sphere. In fact, both Mouzelis and Stones themselves have expressed their affinity toward such efforts. The Greek sociologist argues that conceptual tools should be used "in an effort to move from theoretical compartmentalization and/or mere *juxtaposition* to the effective *articulation* of different perspectives" (2008, 3, stress in original). Stones, on the other hand, is even more precise in his call for bridge-building among compatible theoretical perspectives. "The nature of substantive structuration studies means that they are naturally located at the meso and micro levels [...] and there are many theoretical approaches with a more macro and/or otherwise complementary emphasis that could often be most fruitfully combined with structuration theory. The gain, it seems to me, would be two-way" (2005, 119). In order for a possible *rapprochement* to take place, certain ontological and methodological points have to be clarified. I start by analyzing the ontological and methodological points raised by each theorist and attempt to build bridges whenever possible.

Macro, meso and micro: Actors, structures and levels of analysis

Taking into account the fact that the literature on the micro/macro—actor/structure debate is vast, and that Mouzelis and Stones have each significantly contributed to this highly controversial issue in sociological theory, I will limit the discussion to what these two theorists have argued.

Ontology in Mouzelis

As an overreaction to, mainly Parsonian functionalism and structural Marxism, where structures were omnipresent, and society pulled the strings behind actors' backs, various schools of thought, such as symbolic interactionism (Goffman 1959), phenomenology (Berger and Luckmann 1967), ethnomethodology (Garfinkel 1967), methodological situationalism (Collins 1981) and rational choice theory (Goldthorpe 2000) placed the individual or the interactions among individuals as the locus of analysis. One thing these approaches have in common is the equation of micro with actor and macro with structure. Therefore, a very common practice is the attempt to bridge the micro–macro divide by explicitly or implicitly construing elements at the macro levels as *aggregates* or *configurations* of actors and their interactions. What Mouzelis (1994, 1995, 2008) argues is that this conception offers a very distorted portrayal of society. For him, micro can refer to an actor, interaction or institutional structure the effects of which do not stretch widely in time and space. It can thus refer to an intern in a multinational corporation, a day-to-day interaction among two farmers in a coffee shop or a hunting festival taking place annually in a remote Greek village. At the same time, actors, their interactions and structures can be macro so long as they have consequences felt widely in time and space (Mouzelis 1992, 123). Therefore, macro actors can be heads of companies such as Dow Chemical or GlaxoSmithKline; macro interactions can be the face-to-face interactions among members of EuropaBio; and macro structures can be the global practice of patenting. In using the term macro-actor, Mouzelis refers to *collective actors* (business organizations, trade unions, political parties and so on) and *mega actors* (individuals with high levels of economic, political or cultural capital).

> To ignore such obvious differences and to consider all face-to-face encounters as micro is simply to ignore the hierarchical aspects of complex societies and the fact that actors contribute very unequally to the construction of social orders. In other words, it is to ignore the fact that those who occupy positions with privileged access to the economic, political, and cultural "means of social construction" contribute more significantly to the construction of social phenomena than do those who do not enjoy such access. (Mouzelis 1992, 123)

Methodology in Mouzelis

Regarding the micro/macro distinction, the other issue Mouzelis keeps stressing is that a macro phenomenon should be primarily investigated *horizontally* (that is, by analyzing decisions and actions of macro actors and the constraints/enablements presented by macro structures) and then *vertically*. "This means that institutional structures (system integration) and social actors must be examined *on the same level of analysis*. So, if it is the constitution, repro-duction or transformation of societal *macro*-institutions that is at issue, we must start by looking at *macro*-actors and their modes of connection to such institutions; only then can we move 'down' to consider meso- and micro-actors" (Mouzelis 1994, 117–18, stress in original). It may be worth repeating at this point that the same idea of distinguishing between levels of analysis and recognizing that processes and mechanisms present in each level con-stitute *discrete causal orders in their own right* has also been advocated by Benton (1991, 20) in his analysis of sciences and higher- and lower-level mechanisms. If for example, the issue under investigation is the establishment of the Large Hadron Collider (LHC, also known as the CERN particle accelerator) one has to focus primarily on the team of chief scientists involved, the decisions of 20 EU member states to support it financially and politically, the sym-bolic/cultural meaning that is ascribed to it and so on—in other words move horizontally—and then "move down" and see how these games played at the macro-level enabled/constrained the various groups of scientists and their face-to-face interactions and trace the processes of "translation," as ANT has it, and how specific technoscientific problems were overcome through the process of resistance and accommodation, as Pickering argues. By not doing so, the (until then) scientific, political, cultural and economic environments of action will be either totally ignored or simply be dropped on the surface of a flat laboratory.

Furthermore, it has to be mentioned that whether the interplay between macro, meso and micro levels is reciprocal or highly hierarchized and strat-ified is open to empirical investigation. In some environments, such as an authoritarian society or a bureaucratic organization, games played at the macro level (political or managerial decisions respectively) heavily influence or basically dictate the rules of social games played at the meso and micro levels. Nonetheless, it can be the case that micro-actors can influence macro-structures when, for example, their input and feedback is asked in order for changes to take place. To put it differently, the relative autonomy of sub-levels should not be denied in an aprioristic fashion (downward reductionism[4]) and, at the same time, the superordinate level of analysis should not be merely con-strued as an agglomeration of its subparts (upward reductionism) (Mouzelis 1994, 137). Therefore, what Mouzelis argues is that in hierarchical configura-tional wholes, moving from micro to macro levels of analysis cannot be done via mere aggregation of individual acts, interactions, encounters and so forth (Mouzelis 2008, 260).

It seems to me that Mouzelis's analysis on the *ontology* of *actors* and the unequal way that macro, meso and micro actors contribute to the produc-tion, reproduction and transformation of social life is very useful and should be retained. In fact, Stones is also sympathetic toward such a distinction and acknowledges it in SST (2005, 67).

Ontology in Stones

Issues of ontology pertinent to sociological theory have been extensively discussed by Stones (1996, 2005). He not only accepts Mouzelis's account of macro/mega actors (2005, 67, 134) but also enriches the ontological terrain of social analysis by discussing the concept of "levels of analysis." Stones, however, consistently avoids the use of macro and micro in order to define the various levels of analysis. Rather, he makes the distinction between *ontological abstraction* and *ontological scale.*

Regarding the former, Stones discerns three levels: the abstract ontological level, the meso-level of ontological abstraction and the ontic level (2005, 77).

The abstract ontological level or "ontology-in-general" refers to "abstract ontological concepts produced at the philosophical level" (2005, 38). At this level one encounters the shape or form ontology has at its most abstract and generalized level (2005, 77). Therefore, when concepts such as the duality of structure, internal structures, external structures, agency and so on, are not further elaborated so as to refer to more specific entities pertinent to a particular time, place and social whole, the discussion remains at the top level of abstraction. Ontology-in-general encompasses all the notions about entities that exist in the social world that apply at *all times* and in *all places.* "The abstract concepts are said to be tools to sensitize the researcher in a general, unspecified, way to the kinds of things she might find in the social world" (2005, 76). In that sense, the original version of TAI developed by Mouzelis belongs to this abstract ontological level.

At the other extreme, the ontic level or ontology-in-situ refers to empirically informed, substantive details and specificities of social entities (i.e., internal/external structures, active agency, institutions and so on), which are informed by ontology-in-general and the meso-level (2005, 77). When a researcher is studying social life at this level, she is looking at particular, concrete and situated entities in particular times and places. She has moved, therefore, from the abstract ontology that sketches a very general layout of, say, how an agent's action is generally related to the external environment, to the concrete ways a *particular* actor on that *specific* day decided to make *that* decision. It is obvious that, in order for such an analysis to take place at the ontic level, detailed empirical work is necessary.

The meso-level of ontological abstraction is related to the way in which "it is possible to talk about at least some abstract ontological concepts in terms of scales or relative degrees" (2005, 77). This level is situated between the two extremes of abstract, philosophical concepts, on the one hand, and concrete specificities on the other. Therefore, in one sense, the meso-level of abstraction functions as a stepping stone between abstract, armchair theorizing and exhaustive empirical research. Consequently, the meso-level entails more or less knowledgeability about a range of different objects, more or fewer choices available to an agent-in-situ, external structures that would likely take more or less time to change and so forth.

While the levels of ontological abstraction are related to the more or less informed claims one makes about specific or nonconcrete social entities, the *levels of ontological scale* are quite different. The ontological scale is related to the "scope and scale of a study," to the investigation of "large sweeps of history" or social games that unfold in a very specific place and time (2005, 81). Although, once again, Stones systematically refrains from using the terms macro and micro for the ontological scale—in fact, he mentions macro, meso and micro studies just once throughout *Structuration Theory* (2005, 119)—he nonetheless allows one to infer what exists at the two extremes of the ontological scale.

At the *macro ontological scale,* one can find studies that belong to the genre of (comparative) historical sociology (2005, 135) that are "historically and geographically expansive" (2005, 127) and whose creators in a sense employ broad brush strokes on large canvases (*ibid.*). The focus of such studies is on the "large forces of history and 'conventional' social structures" (2005, 190). Conversely, the *micro ontological scale* appears to refer to studies that "focus on individual agents" (2005, 189–90) and urge the researcher to disclose the hermeneutic and phenomenological dimensions related to an agent-in-situ. In the middle of the two extremes lies the *meso-level of ontological scale.* This level of the ontological scale is inextricably linked with the conceptual cluster of position–practices that was previously developed.

The methodological points that were just briefly touched upon are more extensively discussed in the following sections. For now, it may seem useful to recapitulate the key ontological concepts of Stones and Mouzelis in Table 5.1 so as to start analyzing how the two frameworks can be ontologically articulated.

The arrows in the left-hand column depict that there is always some form of interaction among the macro, meso and micro levels regardless of whether one refers to ontological abstraction, ontological scale or actors. As far as the second category is concerned, it is not too difficult to realize how the scope and scale of a study are affected by events and/or social entities present at the three different scales. Even when a study focuses on the *longue durée*, the researcher will intentionally or unintentionally make reference to position–practice clusters situated at the meso-level or even actions of particular, most likely prominent, situated agents. Conversely, when the focus is on situated

Table 5.1 A typification of actors, levels of ontological scale and abstraction across the macro, meso and micro levels of analysis.

	Levels of Ontological Abstraction (Refers to the degree of abstraction or specificity of social entities)	Levels of Ontological Scale (Refers to the degree of broadness or narrowness of a study's scope and scale)	Ontological Classification of Actors (Refers to the ontological "weight" of social actors)
MACRO	**Ontology-in-General** Abstract/Philosophical concepts that refer to social entities regardless of time and space. These can include notions such as the duality of structure, external/internal structures, conditions of action, notions of enablements and constraint, active agency and so forth.	**Macro Ontological Scale** Studies that are historically and geographically expansive and focus on large forces of history and conventional social structures.	**Macro/Mega Actors** Actors at this ontological level can perform actions with effects that stretch wide in time and space. These can be collective (*macro actors*) or powerful individuals who possess high amounts of political, cultural, economic or symbolic capital (*mega actors*). Macro/Mega actors have the ability to distance themselves from virtual structures for strategic/monitoring reasons. Examples of such actors include EuropaBio, ABC, the CSA, the British PM, Rothamsted Research, Greenpeace and so forth.
MESO	**Meso Level of Ontological Abstraction** Incorporates variations and relative degrees in concepts of the ontology-in-general. These can include more or less knowledgeability about a range of subjects; more or less critical reflection by an agent upon her internal structures; various types of	**Meso Level of Ontological Scale** Studies that are couched either in the "intermediate temporality" of historical processes or on longer periods of time and/or space. In order	**Meso Actors** Meso actors are collective or individual actors who are placed in the middle of social hierarchies. Their actions have a more limited effect on social practices across time and space when compared to macro/mega actors. Taking

enablement and constraint; fewer or greater numbers of consequences of particular actions, and so on. Some examples may include the notion of "novel agricultural techniques," the concept of "global food security," the intentions of a government "to take public opinion into consideration" and so on.

for structural analysis and hermeneutics to take place, the most germane social entities have to be discerned and focused upon. The meso level of abstraction is inextricably linked with position–practice relations.

into account the relative nature of macro/meso/micro and the above examples of macro actors, meso actors can include a British MP, a chemical company, a laboratory, an NGO of national caliber and so on.

MICRO

Ontology-in-Situ/Ontic Level

Details and specificities of concrete and situated social entities. These can include the dispositions of *particular* actors (e.g., a minister's deeply seated liberal convictions) or *specific* structural enablements/constraints present at a defined time/space location (e.g., the presence or absence of opportunities that the EU framework offers to agbiotech corporations that want to export GMOs to EU member states).

Micro Ontological Scale

Studies that focus on agents-in-situ and perform hermeneutic and phenomenological analyses.

Micro Actors

Collective or individual actors who exist at the bottom of the scale of social hierarchies and possess very limited amounts of some forms of capital. Their actions stretch very narrowly in time and space. Micro actors are less likely to distance themselves from the virtual order of structures, which they most often take for granted. Also, their actions are more likely to reproduce social structures rather than transform them.

agents—their motivations, actions, interpretations and so on—these have to be placed within a social context affected by larger social forces: something that was so convincingly argued by Mills (1959). Finally, when one is dealing with configurational wholes, it becomes apparent that social games played at the top of the hierarchy will most of the time influence the social lives of actors placed at the lower end. At the same time, actions of micro-actors can potentially lead to structural change and how social games unfold at a higher stratum of the hierarchy.

Methodology in Stones

Stones (1996) has written extensively on issues of ontology and methodology. Since ontology was analyzed in the previous subsection, the remaining part of this chapter focuses on methodological points pertinent to the theoretical frameworks used in the book.

Degrees of contextualization: Contextualizers and floaters

Stones uses the terms "contextualizer" and "floater" to describe the two basic plans of action a researcher can follow. A study can be characterized by high or low levels of contextualization—that is, more or less rigorous analysis of hermeneutics, meaning attached to actions, power relations and contiguity. The contextualizer strives to be as exhaustive as possible and aims at high levels of contextualization. While contextualizers may embark on a journey of offering as profound an account as possible, they will never be able to grasp fully and, consequently, portray social reality. This is so because exhaustive contextualization may be ontologically real, but is epistemologically utopian (Stones 1996, 75–76). On the other hand, the floater is engaged in an entirely different enterprise. A researcher is more or less forced to be a floater when there are massive gaps in contiguity. Studies that, for whatever reason, acquire a broader and longer perspective make it imperative for the researcher to float over the surface of events as if in a hot-air balloon. This makes them have an extensive, and at the same time shallow, view of social reality. Floaters, can, of course, land every now and again and examine details of social interactions (including those of laymen), but such instances do not last for too long since the balloonist is bound to float upwards and onwards to different horizons (Stones 1996, 77–78).

What the degrees of contextualization help us realize is how closely related ontology and methodology are. The levels of ontological abstraction and scale of a study have a direct impact on the researcher's decision to be more of a contextualizer and less of a floater, or the opposite. The four theoretical

Table 5.2 The levels of ontological abstraction, scale and contextualization that the two frameworks can be applied to.

Theoretical Framework	Levels of Ontological Abstraction	Levels of Ontological Scale	Degrees of Contextualization
Modified TAI	Meso	Meso	Low to Medium (*Floater*)
SST and Technology/ Living Organisms	Micro	Micro	Medium to High (*Contextualizer*)

frameworks have already been mapped in terms of ontology, so we can now start filling in the gaps with the relevant methodological implications. Since the theoretical framework used in the empirical part of the book is an integration of the Modified TAI and the expanded version of SST (SST and Technology), we can simply focus on just the two, rather than the four, theories (Table 5.2).

In terms of contextualization, the Modified TAI remains at relatively low levels. This is so because the researcher is primarily interested in mapping the horizon of Technology, Appropriation and Ideology in rather broad terms. Since the three main constitutive elements of agbiotech, along with their subparts, are not located in a particular social space, but stretch way beyond the time–space locations of their constitutive actors (e.g., a directive from Brussels may have an impact on a Chinese farmer, and a decision by BASF may trigger protests by NGOs a good year later), there are inevitably massive gaps in contiguity. Consequently, a more or less informed mapping of the field can only be achieved by "floating." Using the Modified TAI, one has to take advantage of the methodological "hot-air balloon" that Stones argues for and select at which points of the horizon one will decide to land and focus upon, and which territories will be acknowledged from afar, but left unexplored. The time–space location at which the floater's hot-air balloon is going to land is not only contingent on the places suggested by the theoretical framework (e.g., farms, laboratories, university departments, ministries and so on) but also depends on the presence of macro/mega actors. Mouzelis's stress on the importance of horizontal and vertical hierarchies proves to offer valuable insight regarding the floater's journey. Once this is done, the researcher must start mapping any relations/patterns that emerge among the highest buildings; how one affects the other; what kind of orderly/conflictual relationships they have and so forth (horizontal hierarchies among macro/mega actors). It is at this point, when the places of interest have been located based on the premises of the theoretical

framework and the existence of macro actors, that SST and Technology/
Living Organisms can help the researcher acquire a deeper understanding
of social reality. The degrees of contextualization can now become significantly higher since, inside the configurational wholes that were previously
discerned, the quadripartite cycle of structuration can start to unfold. This
invites the researcher to delve deeper into the relations between actors and
events in his or her quest for knowledge of the agents' internal structures,
their active agency, the conjuncturally specific functions of the technological
artifacts under consideration, the species-life of the living organisms and so
on. The actual *ways* in which knowledge of hierarchies and the structuration
cycle can be obtained is related to the appropriate *methodological bracketing*
used by the researcher.

Methodological bracketing

The notion of methodological bracketing was formulated by Giddens (1984)
and has been further elaborated by Stones on numerous occasions (1991;
1996; 2005; 2012). Methodological brackets constitute a set of conceptual-
practical tools that can offer regulative and selective guidelines that can direct
the researcher to some dimensions of a social object rather than others (Stones
2005, 120). By using methodological bracketing, some dimensions of social
reality will come to the foreground while others will remain in the shadows—
either tentatively or for the whole duration of a research project. Since the
discussion of the various forms of brackets can very easily expand to areas
that lie well beyond the scope of this book, I limit the discussion only to points
relevant to the Modified TAI and the SST and Technology/Living Organisms
frameworks.

The empirical analysis undertaken by the researcher can take two
main forms: Agent's Conduct and Context Analysis (ACA) and Theorist's
Pattern Analysis (TPA). Agent's Conduct Analysis refers to the analysis of
the *hermeneutic* frames of meaning of lay actors (how *they* understand their
inner motivations and dispositions, how *they* more or less reflexively priori-
tize their concerns, how *their* internal structures negotiate with their active
agency) (Stones 1996, 75; 2005, 122). While Agent's Conduct Analysis
invites the researcher to mute the external environment of action and try to
listen to the inner voices coming from the agent herself (her deeply seated
dispositions, generalized schemata of action acquired through prior socialization, current desires, tendencies to improvise and so on), Agent's Context
Analysis, on the other hand, points toward the external environment. This
form of methodological bracketing is appropriate when the researcher is
interested in directing the agent's phenomenology/internal world *outwards*

to the ways in which, and the extent to which, they possess knowledge of the external environment of action (the amount of conjuncturally specific knowledge they possess regarding the existence of position–practice relations, the asymmetrical distribution of power, availability or unavailability of resources and so on) (Stones 1996, 98; 2005, 122). While both forms of ACA try to "listen" to the story from the inside, TPA follows the opposite method. A TPA analysis tells the story from "outside." In this case, while the researcher inevitably engages in interpretation, they perform an analysis *without* any reference to the hermeneutic account of the social phenomena the situated agents have to offer (Stones 1996, 75). TPA can be broken down into ACA's corresponding counterparts. We can, therefore, speak of Theorist's Conduct Analysis and Theorist's Context Analysis. The former refers to instances in which the researcher ascribes to the agent-in-focus intentions, motivations and knowledge based on their own assumptions/ hypotheses; the latter refers to the researcher's structural assessment of various social influences/causations, how certain position–practices are placed across hierarchies, how the external environment of action favors or hinders certain courses of action, constraints and enablements and more (Stones 1996, 106–7; 2005, 122). Having mentioned the two primary forms of methodological bracketing relevant to the theoretical frameworks used in the book, we can now start explaining how they are related to one another.

The very nature of the Modified TAI scheme encourages a TPA kind of analysis. The Modified TAI invites the researcher to float across the field of agribiotechnology and examine its most germane locations. In instances when the hot-air balloon lands on some critical social sites, most likely the researcher will narrate the story from "outside." They will examine the structural conditions (Theorist's Context Analysis) and how actors inside this enabling/constraining environment produce, reproduce and transform their social environment (Theorist's Conduct Analysis). Locating the most important landmarks in the field (macro/mega actors) and sketching the formation of vertical and horizontal social hierarchies constitute the methodological steps that the theorist decides. Nonetheless, this does not exclude the use of ACA. After TPA has been carried out, the researcher may want to examine how situated agents grasped their external environment, and how they interpreted their own internal structures. However, an unrestrained move in this direction would not be very sensible, for two reasons. First, since a, more or less, wide time–space distance has to be traveled, the theorist will most likely not have the necessary time to juxtapose or complement his/her TPA with ACA. Second, even if this is deemed necessary or plausible (i.e., there is enough time and empirical evidence), ACA cannot take place within the Modified TAI in a convincing manner, as this framework does not have at its core the theoretical baggage that highlights the details of social action. Surely, the voices of macro actors or laymen can be heard, but this meso-oriented framework will not be able to use them in a substantial way, as the appropriate tools are unavailable.

Deeper understanding can take place once TPA of the Modified TAI has been established and the salient configurational wholes have been identified and placed in hierarchies. It is at this point that SST and Technology/Living Organisms with its quadripartite cycle of structuration can reveal how and why certain social games are played and, indeed, if they are in fact played in the way the TPA "assumes" or "interprets" from afar. Inside these horizontal and vertical hierarchies, the researcher can apply ACA. On numerous occasions the researcher will want to examine from "inside" why a certain experiment failed, how come Brazilian farmers are more likely to adopt the cultivation of GM seeds than do French farmers, what kinds of social interactions unfold in lobbying meetings in Brussels, just to name a few examples. A convincing answer to such questions should doubtless include the agent's own hermeneutics and structural analysis. While TPA seemed appropriate for the Modified TAI, using the same methodological bracket for SST and Technology/Living Organisms will render the research project simply incapable of piercing through social reality and depicting how actual agents construe social life. For the sake of clarity, what has been argued so far can be depicted in Table 5.3.

By bridging TAI and SST, I have merely argued that so long as this *rapprochement* is sensible this can be one way that may help in grasping, to a fuller extent, the social games played in the sphere of agribiotechnology. Based

Table 5.3 The two frameworks and their respective ontological and methodological implications.

Theoretical Framework	Levels of Ontological Abstraction	Levels of Ontological Scale	Degrees of Contextualization	Methodological Bracketing
Modified TAI	Meso	Meso	Low to Medium (*Floater*)	*Primarily* TPA (*Theorist's Conduct Analysis and Theorist's Context Analysis*)
SST and Technology/ Living Organisms	Micro	Micro	Medium to High (*Contextualizer*)	*Primarily* ACA (*Agent's Conduct Analysis and Agent's Context Analysis*)

primarily on the writings of Merton, Mouzelis and Stones I have argued that a non-deterministic holistic framework that respects the relative autonomy of the technoscientific sphere and the agency/structure distinction may serve as a useful heuristic tool for answering interesting questions about social reality. Being fully aware that the ontological benchmark of exhaustive contextualization is utopian, and that methodological bracketing inevitably leaves some qualities of social entities in the shadows, we can now move to the next chapter of the book to empirically test the Modified TAI/SST and Technology framework in the field of agribiotechnology.

Chapter 6

THE ROTHAMSTED GM WHEAT TRIALS (I): TECHNOLOGY AND APPROPRIATION

Some Preliminary Clarifications

Before revisiting the conjuncture of the Rothamsted GM wheat field trials, I would first like to reiterate the most salient parts developed thus far and afterward offer a brief conceptual overview of the work that is going to follow. Doing so will hopefully make the orientation of the book clearer and the significance of this chapter more evident. Before doing so, however, it needs to be clear from the outset, that the case study stretches across two chapters; this and the one following. Chapter 6, therefore, engages with the Technological and Appropriative aspects of the field trials while Chapter 7 discusses Ideological elements articulated not only during this specific conjuncture but also recurring across the broader debate. This division of the empirical section is implemented so as to make the substantive material more readily available to the reader and, at the same time, signify the gradual expansion of analytical focus from particular events surrounding the field trials to broader social dynamics unfolding across wider stretches of time and space.

Following Chapter 1, which familiarized the reader with the main themes of this book, a variety of theoretical, ontological and methodological issues were raised. The discussion started with the problem of technological determinism and then explored alternative approaches, such as Misa's "meso-level studies," Hughes's "technological momentum" and Jasanoff's idiom of "co-production." Despite their ontological and methodological merits, these three approaches did not convincingly tackle the imbalance between the "social" and the "technological" encountered, not only in technological determinism, but in the broader theoretical avenues of humanism and post-humanism as well. Chapter 3 discussed specific ontological and methodological contributions found in the writings of Merton, Collins, Latour, Callon, Law and Pickering and argued that, despite the breakthrough these studies have offered in STS, they still do not appear to resolve the underlying tension that surrounds the human/non-human relations.

By referring to the work of Benton, it was suggested in Chapter 4 that instead of construing humans and non-humans in terms of either an uncompromising dualism or a hybrid lacking distinct ontological characteristics, one should place various entities on an ontological continuum. In doing so, the researcher is able to grasp the social and natural worlds regarding both shared ontological features and species-specific characteristics. Therefore, by shifting the discussion away from the traditional literature of STS, it was argued that since such studies, in general, have made the *sociological* study of technological artifacts and scientific knowledge possible, there is no apparent reason one should not approach the discipline's sub-field with a set of theoretical and methodological tools that have been used to explore altogether different facets of social life. The two primary schemes that were discussed, modified and brought closer together were Mouzelis's TAI and Stones's SST. By using a variety of examples and theoretical propositions informed by empirical evidence, it was suggested that so long as the researcher is clear about the various ontological levels that the two accommodate and the particular types of methodological bracketing that are appropriate for each unit of analysis, a fruitful rapprochement can be made possible. In effect, this theoretical and methodological articulation between TAI and SST can prove to offer invaluable insight into the field of science and technology as it places interactions within the broader social context and can be sensitive to both social forces of the *longue durée* and the specificities of face-to-face interactions.

This desired flexibility is ascertained in this chapter of the book. By revisiting the 2012 Rothamsted GM wheat trials, I demonstrate the kind of knowledge that can be acquired through the TAI-SST framework. In this respect, while the case study can prove to have merits in its own right by highlighting the particularities of the field trials and placing their overall significance within the GM debate in the UK and the EU, it should be made very clear from the outset that it does not purport to be a case fully worked through empirical elaboration. Rather, its core function is to use the Rothamsted trials as a gateway to the broader conjuncture by demonstrating the heuristic potential of the suggested framework. In this respect, the discussion that is shortly going to follow unfolds along the methodological and ontological axes that were gradually developed throughout the five preceding chapters.

By using the concepts of "floating" and "contextualizing" with respect to the methodological bracketing of ACA and TPA, the exegesis of the wheat trials will traverse across different instances of time and space. At times, it will rest on *macro* levels of ontological scale and follow the significant historical and sociocultural forces in the ways that these influence position practices and situated agents. We see, for example, how transnational conglomerates have persistently tried to influence the GMO authorization process in the

EU or how EU ministers have been apparently very reluctant to vote for the commercialization of GMOs in fear of enduring political and cultural pressures. In other instances, the ontological scale will become *micro* and focus on the particularities of specific agents and the significance of their positions, dispositions, knowledge and actions. One notable example is the case of Professor Moloney, recently appointed director and chief executive of Rothamsted Research, who holds more than three hundred patents in plant biotechnology and whose principal achievement has arguably been the development of GM canola. Highlighting Moloney's human capital and inferred deeply seated dispositions not only adds another layer of understanding to Rothamsted's mission statement, but also brings to the surface the connotation of various position–practice relations formed around the Rothamsted institute. Moloney's achievements, for example, help us draw links of cooperation between Rothamsted and large biotech firms that overtly endorse the GM wheat trials. Conversely, we also observe conflictual relations between Moloney and various GM skeptics who regard the Rothamsted director as the person responsible for the collapse of the organic canola market in Canada.

A wealth of points raised in the first five chapters reappear in the discussion of the case study. Some of them are further enriched while others simply act as reminders of theoretical discussions that took place in previous passages. The human/non-human tension, for instance, which was approached from a theoretical standpoint in the previous chapters, is tackled at the *ontic* level. Quite lengthy passages are devoted to the correspondence among scientists regarding the unpredictability of gene mutations and the unforeseeable environmental effects that genetic modification can have. Empirical evidence is presented that shows the irreversible damage GMOs can cause to ecosystems and portrays the weakness of scientists and experts when attempting to contain the hybridization of crops in the open environment; a reality that, in principle, humanist approaches cannot accommodate. The concept of technological determinism, to give another example, is discerned in many instances ranging from the mission statements of biotech companies (evangelizing a better future) to one of the speeches by Tony Blair on the paramount importance of genetic modification. Further facets of the notion are also explored in Chapter 7 during the discussion of the "nature of Nature" and the "Precautionary Principle." On the other hand, theoretical constructs like actors, position–practice relations, habitus, external environment of action and more are not further elaborated, but are used so as to show the reader how the theoretical framework meshes with the empirical study and the ways the two are intertwined.

As the discussion now starts to unfold, I make the necessary cross-references to the previous parts of the book in order to highlight the aspired cohesion between theory and empirical evidence. Such references are brief when they are more or less self-evident, but quite lengthy when they require fuller explanation. In this way, the rigor, flexibility and openness of the theoretical framework is demonstrated through a case study that has drawn considerable attention by the media and is often associated with the "second push" by the agbiotech firms to bring GMOs to the UK.

Technology

On June 20, 2011, Rothamsted Research (RR) submitted an application (Ref: 11/R8/01) to the Department for Environment, Food and Rural Affairs (DEFRA) requesting permission to conduct a field trial of GM wheat plants on the Rothamsted Farm in 2012–13 in order to test whether GM wheat plants capable of making an aphid pheromone alarm are able to resist aphids under field conditions (Rothamsted Research, 2011b). In turn, DEFRA took advice from the independent Advisory Committee on Releases to the Environment (ACRE) and agreed on terms with the Food Standards Agency (FSA). On September 15, 2011, Lord Henley—by authority of the Secretary of State for DEFRA—granted consent to Rothamsted Research for the release of GM wheat (DEFRA 2011), as it was decided that the file submitted by Rothamsted Research (Rothamsted Research, 2011a) included all the necessary details and complied with the risk-assessment methodology outlined by Directive 2001/18/EC (European Commission, 2001) (Table 6.1).

Authorization of the field trials

According to EU legislation, a person or a company/organization wishing to introduce GMOs into the environment for experimental purposes (e.g., field trials) must first obtain written authorization from the competent national authority of the member state within whose territory the experimental release is to take place. In the UK, the national authority responsible for granting such authorizations is DEFRA. Authorization is given on the basis of an evaluation of the risks to the environment and human health presented by the GMO(s). As stated in Directive 2001/18/EC, the objective of the environmental risk assessment is to identify and evaluate potential adverse effects of the GMO(s) such as direct/indirect, immediate/delayed effects on human health and the environment that may result from the deliberate release or placing on the market of the GMO(s). Furthermore, the environmental risk assessment requires evaluation in terms of how the GMO was developed and examines the potential risks associated with the new gene products produced by the GMO (for example toxic or allergenic proteins), and the possibility of

Table 6.1 Basic information about the GM wheat field trials.

GMO Released	Aim of the experiment	Duration of Field Trial	Location	Funding
• Genetically Modified Cadenza Wheat (*Triticum aestivum*) of the Spring Variety. • The wheat was genetically modified to produce the (*E*)-β-farnesene (EBF) pheromone.	• The volatile chemical EBF is used by aphids under attack as an alarm pheromone or signal and causes other aphids to stop feeding and move away from the source. It also acts as a repellent to colonizing aphid morphs. • The purpose of the trial was to test whether these plants are better able to resist aphids under field conditions. • The experiment was carried out by Rothamsted Research	• The field trials lasted for 5 to 6 months (March/April–August/September) in 2012 and the equivalent time period again in 2013. • As of 31 December 2013, the field work for the experiment had reached completion. • The results of the experiment were published on 25 June 2015.	• Modified wheat was planted in eight 6m × 6m trial plots at Rothamsted's research center in Hertfordshire. • Each plot was separated from each other by 10m and from the edge of the trial by 10 meters of barley (or space) plus a 3m pollen barrier of wheat. • Completely surrounding the site was a 2.4m high chain-link fence (with lockable double gates) to prevent the entry of rabbits and other large mammals including unauthorized humans.	• Funded by the BBSRC • Total costs for the research project were £732,000 • An extra amount of £2,238,439 was spent on fencing and other security measures

Source: Rothamsted Research (2011a).

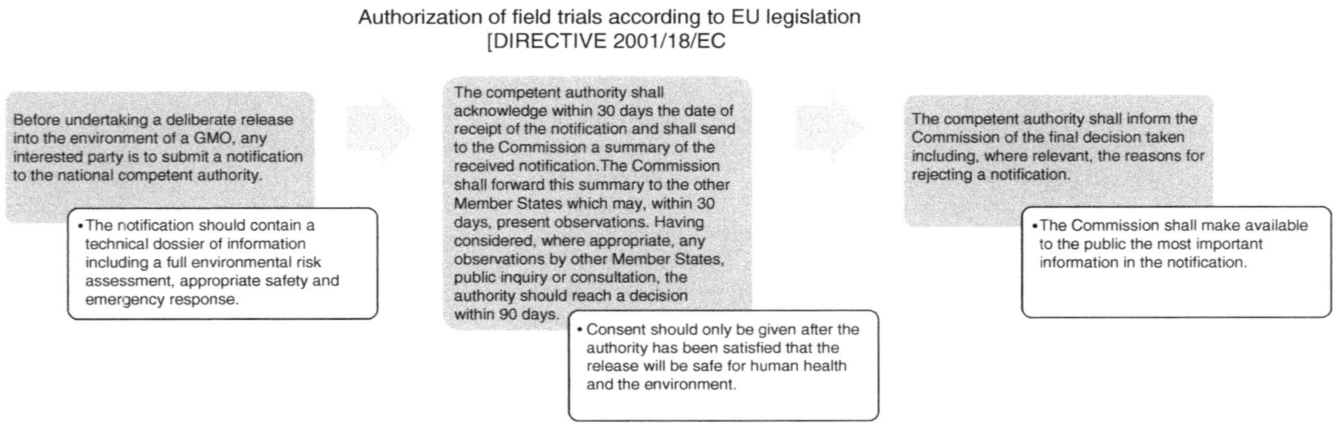

Authorization of field trials according to EU legislation
[DIRECTIVE 2001/18/EC]

Before undertaking a deliberate release into the environment of a GMO, any interested party is to submit a notification to the national competent authority.

- The notification should contain a technical dossier of information including a full environmental risk assessment, appropriate safety and emergency response.

The competent authority shall acknowledge within 30 days the date of receipt of the notification and shall send to the Commission a summary of the received notification. The Commission shall forward this summary to the other Member States which may, within 30 days, present observations. Having considered, where appropriate, any observations by other Member States, public inquiry or consultation, the authority should reach a decision within 90 days.

- Consent should only be given after the authority has been satisfied that the release will be safe for human health and the environment.

The competent authority shall inform the Commission of the final decision taken including, where relevant, the reasons for rejecting a notification.

- The Commission shall make available to the public the most important information in the notification.

Figure 6.1 Authorization of field trials according to EU legislation (DIRECTIVE 2001/18/EC).

gene transfer (for example of antibiotic resistance genes). The risk assessment methodology is as follows:

- Identification of any characteristics of the GMO(s) that may cause adverse effects;
- Evaluation of the potential consequences of each adverse effect;
- Evaluation of the likelihood of the occurrence of each identified potential adverse effect;
- Estimation of the risk posed by each identified characteristic of the GMO(s) application of management strategies for risks resulting from the deliberate release or placing on the market of GMO(s);
- Determination of the overall risk of the GMO(s) (European Commission, 2001).

Finally, although the procedure is a purely national issue—as it is only applicable in the member state where the notification was submitted—the other member states and the European Commission are entitled to make observations that have been examined by the competent national authority (European Commission 2005, 4). Figure 6.1 depicts the procedure that needs to be followed for deliberate release into the environment of GMOs within an EU member state.

Rothamsted Research (internal organization and external environment)

Rothamsted Research is the world's longest-running agricultural research station. With a 170-year history, the center's main areas of research are related to agricultural productivity, crop protection and soil science (Rothamsted Research, 2012d). In the *Science Strategy: 2012–2017* document published by Rothamsted, the center's mission statement is defined as "to perform world-class research to deliver knowledge, innovation, and new practices to increase crop productivity and quality and to develop environmentally sustainable solutions for food and energy production" (Rothamsted Research, 2012d). Rothamsted's vision is to become "a world leading biosciences research center for sustainable agriculture." Research in Rothamsted is congruent with the strategy planned by the Biotechnology and Biological Sciences Research Council (BBSRC) and, at the time of the field trials, was divided into four strategic themes:

- *20:20 Wheat*: Increasing wheat productivity to yield 20 tons per hectare in 20 years.

- *Cropping Carbon*: Optimizing carbon capture by grasslands and perennial energy crops to help underpin the UK's transition to a low-carbon economy.
- *Designing seeds*: Delivering improved health and nutrition through seeds.
- *Delivering Sustainable Systems*: Designing, modeling and assessing sustainable agricultural systems that increase productivity while minimizing environmental impact (Rothamsted Research, 2012d).

Rothamsted's overall scientific agenda is aimed at meeting "the challenge of increasing food and energy production in a more environmentally sustainable way" (BBSRC, 2012b).

Rothamsted Research is an independent charitable company sponsored predominantly by the BBSRC. From the annual income of approximately $42 million, $21 million comes from BBSRC (Rothamsted Research, 2012f), and the remaining amount comes mainly from DEFRA, the European Union and Industry (Rothamsted Research, 2012b). BBSRC also invests in equipment and facilities and is the employer of most of the staff at Rothamsted Research. However, since October 2011, all new staff are institute employees. Land and buildings of Rothamsted Research are owned by Lawes Agricultural Trust (LAT). "As such, the entirety of 'Rothamsted' is best thought of as an enduring partnership between three parties with coincident interests: LAT Company Limited, BBSRC and Rothamsted Research and these three elements are combined in and sustained by the Members of the Rothamsted Research charity, thereby facilitating the long-term operational functionality of this association" (Rothamsted Research, 2012d). The institute employs over three hundred research staff, who are "keen to see their work being applied to meet existing and future industry needs and many of the projects carried out by [the] researchers are fully-funded or co-funded by industry or supported with in-kind contribution from companies" (Rothamsted Research, 2012a). The portfolio of industrial partners includes global agribiotechnology corporations such as Syngenta, Dow Agrosciences, Bayer Agriculture, BASF and Monsanto; multinational organizations such as British Sugar and Novozymes Biologicals, Inc., but also small and medium-sized enterprises (ibid.).

BBSRC, the main source of Rothamsted's income, was established by Royal Charter in 1994 by incorporation of the former Agricultural and Food Research Council (AFRC) with the biotechnology and biological sciences programs of the former Science and Engineering Research Council (SERC). BBSRC is a non-departmental public body that reports its activities to and is funded by the UK government through the Department of Business Innovation and Skills (BIS). BBSRC's budget for 2011–12 was around $710 million, and around sixteen hundred scientists and two thousand research students in universities and institutes across the UK were supported (BBSRC, 2012b). The top-level decision-making body of BBSRC is the "Council," which is account-able to Parliament for the activities of BBSRC. At the time the field trials started, the register of the BBSRC Council included members employed by academic institutes and large biotechnology corporations such as Dow Agro Science and Syngenta (BBSRC, 2012a).

At the time of the field trials Rothamsted Research was governed by a board of 14 non-executive trustee directors and chaired by Professor Nick Talbot.[1] BBSRC and LAT each nominate one trustee. The chair is jointly appointed by BBSRC and LAT. All board vacancies are advertised in the national press; the positions are renewable, four-year appointments, and are unremunerated. The chief executive of Rothamsted Research is the institute director, who reports to the board chair and is in attendance at board meetings along with the associate directors. In 2012, the institute director was Maurice Moloney.[2]

Moloney and Pickett: Dispositions, positions and locations in wider networks

The presence of Moloney as the institute director of Rothamsted, together with the presence of Baulcombe and David Lawrence in the Council of BBSRC (affiliated with Dow Agro and Syngenta, respectively), became a point of concern, to put it mildly, among some GM skeptics regarding the overall orientation of Rothamsted (Take the Flour Back, 2012a). Although the exist-ing friction between Rothamsted and some NGOs is examined in the section on Appropriation, it is worth mentioning very briefly the past achievements of Moloney, and how his position at the helm of Rothamsted helped extend and consolidate the position–practice relations beyond the institutes of close affiliation.

On January 14, 2010, BBSRC and LAT appointed Moloney as director and chief executive of Rothamsted Research. Moloney's appointment at the helm of Rothamsted was met with profound excitement from his colleagues. Talbot, then chair of the Rothamsted board of directors, welcomed the appointment, saying: "Maurice Moloney will bring new ideas and leadership to Rothamsted. He has an excellent track record of making fundamental scientific discoveries and applying these to produce new crops and tools for farmers. I look forward to working with him in the future." Acting director at Rothamsted Research, Professor Peter Shewry, also welcomed the news: "Maurice's leadership will not only ensure our research is of the highest international quality but also build on our strength in ensuring it is translated into useful outcomes for the industry and the public" (BBSRC 2010).

Moloney is regarded as a leading authority on plant-cell biology, especially seed biology and its biotechnological applications in crop improvement. He holds more than 300 patents in plant biotechnology worldwide. The director of Rothamsted was the founder of SemBioSys and served as the company's president from 1994–98 and then as chief scientific officer from 1998–2010. Moloney also has significant experience of research policymaking, having served on the National Sciences and Engineering Research Council of Canada 2002–08. Prior to founding SemBioSys, he spent seven years as an academic conducting research on seed-specific gene expression, herbicide resistance and the plant cell cycle. Previously, he was the head of the Cell Biology Group at Calgene, Inc. (acquired by Monsanto in 1997), where he developed the first transgenic oilseed plants using canola as the model. This resulted in a landmark patent in plant biotechnology and eventually became the basis of RoundUp Ready and Liberty Link Canola, which currently commands 85 percent of the canola acreage in Canada (Rothamsted Research, 2012c).

In an article of his, published in 1995, one can discern Moloney's deeply embedded dispositions, which reveal a profound faith in science and technology.

> Agricultural biotechnology and the specific phenomenon of transgenic crop plants has, not surprisingly, received much scrutiny in recent years. It is normal to greet any new technology with a critical, analytical gaze. From the earliest times any technology which represents a paradigm-shift has elicited a knee-jerk reaction of suspicion and concern. It is the job of those involved in such science not only to generate the technology, but also to provide factual information which will enable society to make rational decisions about its desirability. Rationality has not always been the hallmark of the biotechnology debate. Much effort has been wasted on the criticism of biotechnology using arguments that come from narrow philosophical belief systems such as creationism, left or right-wing politics or aesthetic rather than rational views of nature […] Despite our sophistication as a society, we live in an illogical world. Plant genetic engineering has been targeted by some as being generally undesirable without any real attention to its possible effects. Yet we do not question the sale of naturally-occurring plants or plant products. Mustard and horseradish with thiocyanates, cassava which produces cyanides and even rhubarb which makes toxic oxalates could never have been registered as food crops in today's regulatory climate. In forming judgments about the utility of genetically engineered crops, we certainly require a rigorous regulatory system. However, we must recognize that new technology may offer advantages which go counter to our preconceived notions. In a climate of calm, deliberate and rational analysis, we shall discover that it would be foolish to deny ourselves or our environment the advantages which may be offered by plant genetic engineering and allied technologies. (Moloney 1995)

The then-professor at the University of Calgary seems to be espousing the second facet of technological determinism (i.e., the idea that as technology evolves, social and cultural structures must/should adapt accordingly). Nonetheless, it is of minor importance whether or not Moloney espoused technological determinism by evangelizing a society that would not be "foolish to deny itself" the benefits of biotechnology. What should be retained kept from his writings, however, is his apparent deeply couched tendency to marginalize any social, political or cultural considerations that may hamper the advent of biotechnological artifacts as elements of everyday reality.

Moloney's greatest achievements—the development of RoundUp Ready and Liberty Link Canola—have received fierce criticism from a vast number of individuals concerned with agribiotechnology. The significant issue with RoundUp Ready GM canola is that it has "contaminated seed stocks, farm fields, and the canola distribution systems to such an extent that it has virtually wiped out the organic canola market for Saskatchewan farmers" (D. Olson 2005, 155). In fact, the extent of contamination in Canada is such that 90 percent of certified non-GM canola seed samples contain GM material (Friesen, Nelson and Van Acker 2003). The extent of such contamination comes as no surprise, as a study conducted by DEFRA and the Scottish Crop Research Institute found that GM canola can cross-pollinate with non-GM canola more than 26 km away (Ramsay, Thompson and Squire 2003).

With the successful completion of the field trials in December 2013, Moloney departed from Rothamsted Research as he took a new post as group executive for the Commonwealth Scientific and Industrial Research Organization (CSIRO). Moloney's four-year leadership at Rothamsted proved to be a huge success. Among his accomplishments was securing a multi-million-dollar pound research partnership with Syngenta, expanding the institute's public funding by doubling BBSRC's strategic investment and obtaining several significant awards from the EU. The BBSRC also offered more than $17 million in infrastructure investment for new building projects, including the industry collaboration hub and a new conference center. Professor Doug Kell, chief executive of BBSRC, praised Moloney for his work at Rothamsted: "Maurice has made huge contributions to UK bioscience over the last four years, not only for Rothamsted Research but for the wider BBSRC and scientific community. Thanks to his leadership, the profile and impact of the Institute's excellent research has been greatly enhanced. Maurice recognized the value of Rothamsted's research—not only for improving our knowledge but also for delivering economic and social benefits for the UK—and built on this for the benefit of the Institute with exceptional results. This has ensured an excellent future for Rothamsted Research and for UK agri-science" (qtd. in Rothamsted Research 2013).

The GM wheat field trials were conducted under the scientific theme of 20:20 Wheat; one of the four key programs of Rothamsted mentioned previously. The principal investigator of this research program was Professor John Pickett, and the leader was Dr. Toby Bruce. Other researchers involved in the aphid-repellent GM wheat experiment included Professor Huw Jones, Janet Martin, Professor Johnathan Napier, and Lesley Smart. For this research, Rothamsted Research collaborated with IDna Genetics Ltd[3] (UK) and GenScript Corporation[4] (USA) (Rothamsted Research, 2012d).

The presence of Pickett as the principal investigator of the program was yet another cause of discontent among some NGOs. This is so because Pickett was involved in the "Pusztai Affair" more than a decade previous. In 1997, Dr. Arpad Pusztai and his colleagues at the Rowett Research Institute discovered rats that consumed potatoes genetically modified with the GNA lectin suffered from totally unexpected changes in the size and weight of their body organs. When Pusztai and Stanley Ewen submitted their paper to *The Lancet* for publication, the editor of the journal, Dr. Richard Horton, sent the article to six reviewers and Pickett was one of them. While there are some suspicions that Pickett had not even properly read Pusztai's and Ewen's study,[5] when he became aware of the journal's intentions to publish it, he was so outraged that he decided to go public, breaking the scientific rule that reviewers should remain anonymous (Rowell 2003, 116). After taking his concerns to the Royal Society—which informed him that they were already preparing a press release—Pickett initiated a spoiler article at *The Independent* newspaper under the headline 'Scientists revolt at publication of flawed GM study" and alleging the paper had supposedly failed the peer-review process, yet *The Lancet* nonetheless was about to publish it (Brian 2012). Pickett claimed: "It is a very sad day when a very distinguished journal of this kind sees fit to go against senior reviewers" (qtd. in Connor 1999). The information spread fast. The chief executive of the BBSRC, Professor Ray Baker, claimed, "It is irresponsible for *The Lancet* to publish a paper which has been deemed unworthy of publication by referees" (qtd. in Rowell 2003, 116). Steven Cox, the Royal Society's executive secretary, said, "If they publish without any disclaimer or without making clear the reservations of the reviewers then we would take a very serious view of it" (qtd. in Connor 1999).

By stating that other reviewers were opposed to publication, however, Pickett had clearly lied. According to Horton, the editor of *The Lancet*, "A clear majority of *The Lancet*'s reviewers were in favor [...] Of the five technical experts four were in favor of publication, but one was firmly against, one was in favor of publication, but felt it was flawed, the others were in favor for its scientific merit" (qtd. in Rowell 2003, 116).

Apart from scientific public outcry, Pickett's unfounded allegations triggered less overt but more persistent pressures towards the journal editor. Horton said he was threatened by a senior member of the Royal Society that his job would be at risk if he published the controversial research. While he declined to name the one who telephoned him, *The Guardian* identified him as Professor Peter Lachmann, then vice-president and biological secretary of the Royal Society, president of the Academy of Medical Sciences and professor of immunology at Cambridge University. Horton claimed that he was called "immoral" for publishing Pusztai's and Ewen's paper while he "knew [it] to be untrue" and if he, nevertheless, went on to publish it this would "have implications for his personal position" as editor. Lachmann, however, "categorically" denied making any threat to Horton during the call. "This is absolute rubbish, it would never have crossed my mind […] I didn't accuse him of being immoral. I said there were moral difficulties about publishing bad science. I think I probably suggested to him that he knew the science was very bad […] I don't think I used the word untrue" (qtd. in Flynn and Sean 1999).

After his influential[6] post at the Royal Society, Lachmann is currently on the advisory board of Sense about Science, a charitable trust that promotes the public understanding of science and which was actively involved in the Rothamsted field trials.

GM wheat: Issues of discord

The organisms were planted on March 22, 2012, at Rothamsted's research center in Hertfordshire were wheat plants that had been genetically modified to test a novel resistance to aphids, a major pest of cereals. The genetically modified plants were made by inserting new DNA into the wheat genome using microprojectile bombardment and tissue culture. The two new genes were synthetic, that is, they were not taken from another organism but were chemically synthesized to function like wheat genes. The proteins they encoded are common in nature, and the particular forms used are similar to those found in *peppermint* and *cows*.[7]

The new genes added encode enzymes that lead to the production of a volatile chemical that is naturally produced by aphids and many other plants. The volatile sesquiterpene (E)-β-farnesene (EBF) is used by aphids under attack as an alarm pheromone, or signal, and causes other aphids to stop feeding and move away from the source. It also acts as a repellent to colonizing aphid morphs. In addition, emission of EBF was expected to cause increased foraging by predators and aphid parasitoids.

The plants also contained two selectable marker genes, both of which originate in bacteria. The bar gene gives the plant resistance to glufosinate herbicides and was used in the selection of transgenic plants. The nptI gene confers resistance to the antibiotic kanamycin and was used in the gene cloning steps. Glufosinate was not used to control weeds on the trial site, and it was not considered harmful in the context of this trial (Rothamsted Research 2011a and 2011b).

These scientific practices, however, raised serious concerns among some scientists outside Rothamsted who questioned the experiment's effectiveness and overall safety. In August 2011, molecular geneticist and developmental biologist Dr. Ricarda Steinbrecher, codirector of the EcoNexus research organization, submitted a report to DEFRA in which the major points of Rothamsted's application (Ref. 11/R8/01)—highlighted in the above paragraphs—were brought into question. I focus on three main areas of concern raised by Steinbrecher, but which are also supported by many other scientists as well.

Firstly, Steinbrecher questioned the underlying hypothesis of the Rothamsted team (see Table 6.1). To repeat, Pickett and his team wanted to perform field trials in order to test whether the EBF pheromone produced by the wheat would repel aphids as successfully as they did under laboratory conditions. Steinbrecher, on the other hand, argued that the hypothesis was neither developed nor robust enough to warrant an outdoor step that introduced more variables and risks (Steinbrecher 2011, 2). While the data from initial indoor experiments carried out by Rothamsted had not been made available to Steinbrecher, some other findings published by independent research teams are of great interest.

Before proceeding with the genetic modification of wheat with the EBF synthase gene, Rothamsted modified thale cress (*Arabidopsis thaliana*) with the EBFS gene (derived from peppermint) and provided other research teams with this GM *A. thaliana*, who used it in experiments with the green peach aphid (*Myzus persicae*), known to produce and respond to EBF. Two research teams found the plant-produced EBF was not functioning as was intended and did not offer the GM plant protection from aphids. Martin De Vos et al. (2010) found that exposure to EBF led to "habituation within three generations" (14673) and, therefore, the plant-based production of EBF did not appear to offer any agricultural benefit. Furthermore, Grit Kunert et al. (2010) state that "no evidence was found for the ability of EBF to directly defend the plant against aphids. EBF emission did not significantly repel winged or wingless morphs from settling on plants. Nor did EBF reduce aphid performance, measured as reproduction, or lead to an increase in the proportion of winged offspring" (1). Following Kunert et al. (2010) and Dirk Avé et al. (1987), Steinbrecher argued that although aphids emit EBF alarm pheromone in a *pulsed fashion*, the EBF produced by plants will be present *constantly* due to the constitutive expression of the EBFS gene. "The difference in the way that

EBF is emitted is crucial as it may function as an important cue in informing aphids whether the EBF is coming from attacked conspecifics (so it is necessary to take evasive action) or from a plant (so there is no immediate predation risk)" (Kunert, Reinhold and Gershenzon 2010, 10). Therefore, Steinbrecher argues, the hypothesis forwarded by Rothamsted, if anything, is not supported by enough data to justify environmental releases. Findings of other researchers point to the fact that long-term indoor greenhouse trials have to be carried out so that variables can be minimized and assumptions safely tested (Steinbrecher 2011, 3).

The second line of concern raised by the codirector of EcoNexus is that data, information, and risk-assessment deliberation are insufficient regarding the molecular and phenotypic characterization of the GM wheat. As was previously mentioned, the wheat was modified via the procedures of particle bombardment and tissue culture. Both these transformation procedures cause a large number of mutations (genome-wide and insertion-site mutations), which in turn can lead to unintended effects and consequences. "Any transformation-induced mutation which affects functional DNA sequences has the potential to result in unexpected phenotypic consequences. This is true for single base pair changes and for large deletions and rearrangements. Thus, in a commercial crop plant, every transformation-induced mutation is a potential hazard" (Wilson, Latham and Steinbrecher 2006, 220). While the assessment of the unpredicted consequences/effects that any such mutations may trigger "should be part of any risk assessment of GMOs intended for environmental release, whether for field trials or commercial release" (Steinbrecher 2011, 3), it appears that this is not entirely the case with the Rothamsted wheat trials. More specifically, in the application of the research center it is clearly stated that:

> We have not analyzed the position or the structure of the insertion nor sequenced the flanking genomic DNA. Apart from the expected phenotype of EBF emission, these plants are indistinguishable from untransformed controls. No other changes to the plant morphology or development are apparent. (Rothamsted Research, 2011a, 7)

Since the position, the structure of insertion, and sequence of the flanking genomic DNA had not been analyzed, the claim that the GM plants are "indistinguishable" from the non-GM plants was not based on experimental data, but rather constitute a broad assumption or merely an opinion. The absence of concrete data led Steinbrecher to conclude:

> Risk assessment is intended to protect the environment as well as human and animal health from negative impacts and potential harm. Without

data, such assessment becomes guesswork, which cannot provide the degree of certainty required for decision making. Lack of relevant data requires the application of the precautionary principle or precautionary approach until such data is made available and assessed. (Steinbrecher 2011, 4)

Inadequate information is also observed, Steinbrecher underlines, in the characterization of the two inserted genes (those mentioned above that resemble genes found in peppermint and cows) as "synthetic." It is not only the fact that the very definition of a gene as "synthetic" is arbitrary and would mean different things to different people, but also, Rothamsted's account of these two genes seem to be contradictory. More precisely, in Rothamsted's "Application for consent to release a GMO—Higher Plants"/ Part B, the applicants state: "The two new genes are synthetic i.e. they were not taken from another organism but chemically synthesized to function like wheat genes. The proteins they encode are common in nature and the particular forms used here are similar to those found in peppermint and cow" (Rothamsted Research, 2011b, 1). This statement, however, is not compatible with how Rothamsted Research described the genes to BBSRC when they applied for their grant (Ref BB/G004781/1). In that document, the inserted gene is not identified as "synthetic," but the reader is given the impression that the experiments have been carried out by the use of the gene found in peppermint plants.

> [W]e have isolated the gene responsible for the production of pure aphid alarm pheromone in peppermint plants. By inserting this gene into other plants we can make them produce the pheromone and we have recently performed this transformation with a simple plant called thale cress which is widely used by biologists as a model. When we did this, aphid pests were no longer attracted to the transformed cress. Furthermore, natural insect control was improved because a key natural enemy of aphids, called a parasitoid, searched for aphids for a longer period of time on the transformed thale cress. Following this success, we now need to carry out a similar transformation with wheat, a crop plant of worldwide importance, so as eventually to exploit this system for aphid pest control and breed a new generation of environmentally friendly GM crops. (Rothamsted Research 2008, 2)

It is therefore not clear why the genes are called synthetic, what this characterization entails or in what ways the EBFS and FPPS genes differ from the original peppermint and cow sequences, respectively. Following the lack of information provided by Rothamsted, Steinbrecher argues that the origin of the gene sequences should be provided, as it is relevant to potential instances of gene silencing, horizontal gene transfer and viral insertion sites (Steinbrecher 2011, 3–4).

The third main point of concern expressed by the codirector of EcoNexus was the presence of antibiotic resistance marker genes. Marker genes constitute an integral part of the GM process and are used in order to help the researcher identify which cells have been successfully modified and which have not. The use of *antibiotic resistant* marker genes is a very common, but highly controversial, practice. "GM cells are given an antibiotic resistance gene, so when they are grown in a medium containing the relevant antibiotic, only the cells that have been successfully transformed will survive. However, the worry is that these genes might be transferred to bacteria in the guts of animals or humans, and that diseases could become resistant to antibiotics" (Rowell 2003, 126). In fact, the British Medical Association (BMA) believes that "the use of antibiotic-resistant marker genes in GM foodstuffs is a completely unacceptable risk, however slight, to human health" (qtd. in Diamand 2001, 17). The GM wheat produced by Rothamsted contained two marker genes; the *bar* gene and the *nptl* gene. The latter confers resistance to the antibiotic kanamycin[8] and others used in the fight against serious bacterial pathogens. As Rothamsted acknowledges in its application: "Some aminoglycoside antibiotics including kanamycin are considered important for clinical treatment, especially for second line treatment for multi-resistant tuberculosis (kanamycin) and in gut irrigation in, for example, encephalopathy (neomycin)" (Rothamsted Research, 2011a, 21).

"The fact that the nptl gene can also be found in the natural environment is no excuse for its presence in the GM wheat" argues Steinbrecher, who believes the antibiotic resistant marker gene should be removed prior to any release (Steinbrecher 2011, 5).

Summary

In the first part of the TAI/SST scheme, the major issues related to Technology were discussed. So far, I have sketched out the major social actors (collective and individual), the rules and resources, and the position-practice networks that constitute the external environment of action of the Rothamsted Research team responsible for the design and field trials of the GM wheat. Depending on the aim of the empirical part and the data available, some aspects of the external environment were assessed quite thoroughly (funding, the center's mission statement, who the macro actors are and what these signify and so on) while others (how, for example, the habitus-informed character of Pickett or Moloney's conjuncturally specific knowledge of his own external environment shape the way research is construed and conducted) were touched upon. It has been clearly stated that Rothamsted Research, which is mainly publicly funded via the BBSRC, aspires to play a world-leading role in the future of agribiotechnology. The GM wheat field trials constitute a step in this direction and were a part of the general 20:20 Wheat strategic theme. Moving away from the external environment of action to the actual technological artifacts, we saw that the functionality and properties of the GM wheat were far from undisputed. Among many scientists who found the methods and technology used in the specific experiment highly controversial, Steinbrecher made her concerns regarding human health and the environment known to DEFRA and explicitly expressed her objection to the environmental release. The decision of DEFRA to allow the field trials to take place can be explained by assessing the social games played by collective and individual actors as they vie to appropriate Technology.

Appropriation

The second facet of TAI shifts attention away from the forces of production related to the specific technology and brings to the forefront the various social actors that struggle to appropriate/control the technological artifact. It seems that in a situation where a GMO has received approval and is cultivated in a country for food/feed, Appropriation can be witnessed in three major areas: R&D (forces pertinent to the creation of the GMO and actors who object to it); farming (how the cultivation of the GMO is structured (i.e., who sells the GMO to the farmers and under what conditions, who owns the seeds, legislation regarding the coexistence of GM and non-GM crops and more); and distribution (who supported/opposed authorization and on what grounds, whether the GM food/feed was adopted/boycotted and for what reasons, and so on). In the case of the Rothamsted GM wheat, Appropriation appeared to unfold primarily in R&D, as the GMO was only in the process of testing. Nonetheless, it seems to be the case that some elementary forms of Appropriation existed in Farming and Distribution as well. Regarding Farming there appeared to be some reactions from NGOs and farmers' unions regarding the potential adoption of GM wheat. As for Distribution, a respectable number of actors seemed to be stressing the enormous positive impact that GMOs may have on the British economy and favor much more satisfactory European legislation that would allow the cultivation of GM crops. Since the notion of potential cultivation/farming was used either as a reality of the near future or as the next logical step following the field trials and was more or less construed as part and parcel of the R&D phase, I devote one section for the two. Also, although Appropriation in Distribution for the GM wheat per se did not exist, there were—and still are—many struggles for control, and the outcomes would affect the authorization of the GMO under discussion had the field trials been successful.

R&D and farming

As mentioned in Chapter 1, a number of collective and individual actors expressed their concern about the safety and overall implications of these field trials. The "GM Wheat? *No Thanks!*" campaign was launched precisely to stop Rothamsted Research from releasing GM wheat into the British environment. The supporters of the campaign, which included among others GeneWatch UK, Greenpeace, EcoNexus, the Gaia Foundation, Real Bread Campaign, Bakers Food and Allied Workers Union and many others, used institutional means in their effort to control the course of the field trials. These actors wanted to raise public awareness of what was at stake and, in this way, "send a strong message to the Government and regulators" that GM wheat (trials) should be banned (GM Freeze 2012c). Another network of individuals, however, the grassroots movement Take the Flour Back, opted for non-institutional, violent means to halt the experiment, which they construed as a "real, serious and imminent contamination threat to the local environment and the UK wheat industry" (Take the Flour Back 2012). Despite constant efforts from Rothamsted Research to engage with Take the Flour Back in public dialogue so as to openly discuss apparent issues of concern, the grassroots network declined to participate in a central debate chaired by George Monbiot. Instead, Rothamsted Research and the UK government were given an ultimatum to "remove the GM plants themselves" by May 27, 2012. Otherwise, the grassroots network would follow the "only avenue open" to them, which was to "decontaminate" the area (Take the Flour Back 2012). Amidst fears of vandalism, scientists from Rothamsted Research launched the "Don't Destroy Research" appeal and petition, which constituted a plea from scientists and individuals from all over the world to make activists reevaluate their offensive stance. More than six thousand individuals signed the petition (Sense about Science 2012) and some 44,000 people became aware of Rothamsted's appeal by watching the video uploaded by the institute (Rothamsted Research 2012h). On the planned day of "decontamination," the police managed to stop protesters from reaching the site of the field trials and, apart from two arrests, "no disturbances in the area" were recorded (BBC 2012).

During this discursive struggle to safeguard their right—granted by DEFRA—to pursue the field trials, Rothamsted was not alone. The charitable trust Sense about Science endorsed Rothamsted's Don't Destroy Research appeal by promoting the campaign and inviting Rothamsted researchers to answer the public's inquiries about GMOs (Sense About Science 2012). During the short debate that took place on May 17 (i.e. ten days before the scheduled protests) on BBC's "Newsnight," next to Pickett was the managing director of Sense about Science, Tracey Brown, the two of them promoting, in tandem, a pro-GM stance, while criticizing Jyoti Fernandes, a member of Take the Flour Back, for the planned "decontamination" of the sight ("Newsnight" 2012). On the day of the protest, Theo Simon from the protest group appeared in a debate against Graham Jellis (special professor in applied plant pathology at the University of Nottingham, member of the British Crop Production Council, the BBSRC Sustainable Agriculture Strategy Panel, director of R&D at HGCA, and member of the Rothamsted Research Association) (BSPP 2012). Shortly before the debate began, a brief broadcast from the field at Rothamsted showed Moloney commenting on the protestors: "We haven't been able to get them to engage us, quite honestly. There's a lot of rhetoric; we don't even know who they are. Anybody who doesn't want an answer to a scientific question is really missing the point" (Channel 4, 2012). While the debate between Simon and Jellis revolved around facts and figures of GMOs, the issue of the protest was once again stressed as the program ended with Jellis stating: "We need to talk rather than protest" (Channel 4, 2012).

On the other hand, GM Freeze and other NGOs had explicitly stated their "long-standing frustrations with the way a variety of issues related to GM food and crops are reported by the media, including the understanding of the science, the motivations of those questioning the usual view and the politics of Big Ag." The concerns expressed did not relate to the amount of time offered to each side, but rather to "the quality of discourse and the failure of the media (in general, there are exceptions) to engage in an exposure of the deeper questions, ethical and scientific, in favour of the easier option of portraying well-founded concern and questioning as 'fear' of change or technology" (GM Freeze 2012a).

As the field trials were almost always presented in the media in light of the planned "decontamination," members of some of the most prestigious public relations firms interpreted the altercation between Rothamsted and Take the Flour Back as a "communications battle," where the research center came out as a clear winner. Nick Laitner, director of public affairs at MHP Communications, applauded Rothamsted's handling of the media, saying: "The scientific community and authorities have done well to portray the protesters as extremist zealots." In the same line of thought, Alex Deane, head of public affairs at Weber Shandwick, said: "It's great to see a research institution handle comms so well. If the protesters are happy to shout slogans, but not debate in a rational way, what does it say about them and their cause? [...] Protesters in the UK have been getting worse and worse. These cases will continue, so researchers are going to have to get into the comms game and handle things well—like Rothamsted has. Science one, whining greenies nil" (Luker 2012).

Further support came from Peter Kendall, president of the National Farmers Union (NFU—the largest farmers' organization in England and Wales), who likened anti-GM activists to Nazis. In a speech to MPs in the House of Commons to launch the new NFU "Farming Delivers for Britain" campaign, Kendall compared the random damage caused by a single individual—not related to Take the Flour Back—to the Nazi book burners of the 1930s. "I have to condemn the scandalous attempts over the weekend to destroy the trials of GM wheat at Rothamsted. This is criminal, and must be dealt with as such. It's worse than that. It is the willful imposition of ignorance, directly comparable to Nazi book-burning in the 1930s [...] Those who have incited this activity, under the guise of a peaceful demonstration, should hang their heads in shame" (Driver 2012).

Though their suggested plan of action drew considerable attention, Take the Flour Back was not only about decontamination. In fact, the concerns raised by the network were either identical or overlapping with those expressed by GM Freeze in their letter to DEFRA, where Rothamsted's application was scrutinized on scientific grounds (Riley 2012). Although their criticisms were targeted at the Rothamsted field trials, they tended to expand well beyond the experiments carried out at Hertfordshire and expressed overall concerns about GMOs.

For the purpose of this section, I will only refer to the issues forwarded by GM Freeze and Take the Flour Back related to the GM wheat under discussion. Putting aside the scientific issues raised by EcoNexus and GM Freeze (including health and environmental safety concerns), which were mentioned in the previous section, the main lines of criticism launched by social actors who have attempted to appropriate/control the GM wheat by not allowing its testing were the following: (a) the field trials are unnecessary; (b) they serve interests of transnational corporations.

Potentially useful or certainly pointless?

To recapitulate, Rothamsted had been trying to develop "novel ecological solutions" in order to overcome the problems caused by aphids on wheat. As cereal aphids reduce yields by sucking sap from plants and transmitting the yellow dwarf virus, a large number of UK farmers try to control the aphid population with the use of chemical insecticides. However, repeated use of insecticides often leads to resistant aphids and kills other non-target insect species, including the natural enemies of aphids, which could have a further impact on biodiversity. One potential solution to this problem suggested by Rothamsted has been to develop a GM wheat that produces high levels of an aphid-repelling odor, or alarm pheromone, which aphids themselves produce to alert one another to danger. This odor, (E)-β-farnesene, is also produced by some plants as a natural defense mechanism and not only repels aphids but also attracts the natural enemies of aphids, such as ladybirds. If the field trials proved to be successful, this particular technology would help promote sustainable and environmentally friendly agriculture (Rothamsted Research, 2012e). On the day of protests against the field trials, Moloney was on the site and once again expressed his support and ambitions regarding the GM wheat by telling Channel 4: "Most of the scientists at Rothamsted would say we are the greenest people on the planet. Here we have an opportunity, it's not the only way to do it, but an opportunity to replace the use of insecticides on the wheat crop, if it works […] Most people would say that's a great idea if we can reduce the use of any kind of pesticide" (Channel 4, 2012).

For protesters, however, the field trials were at best pointless or, for some, even extremely dangerous. First, the spring wheat variety modified by Rothamsted was deemed as an inappropriate choice, as spring wheat only accounts for 1 percent of the UK crop. What is more, severe aphid infestations on spring wheat are infrequent, and even when they do occur, the control costs to farmers can be as low as eight dollars per hectare (Lundgren 2012). Rothamsted acknowledged that "winter wheat would have been a better model for UK agriculture but this is an early stage experiment and 'spring type' wheat provides a more efficient experimental system to test our hypothesis […], and, therefore, the experiment can be conducted in a shorter time frame […] Our 'spring type' wheat, Cadenza, also has good frost-tolerance which will protect the experiment from our unpredictable UK weather" (Rothamsted Research, 2012f). The same issue was brought up by Lawrence Woodward— representing "Citizens Concerned about GM," a group of people not necessarily opposed to GM—during the "Newsnight" GM debate on BBC when he persistently reminded Pickett that: "This trial isn't necessary, it's on a crop that really doesn't suffer from aphid problems […] aphid is not a problem in spring wheat […] I can't see a justification for that […] in the last six seasons there have been no aphid problems in the UK." The principal investigator of the Rothamsted team replied: "We can work on winter wheat[; …] winter wheat would be a very nice model, but it would be a lot more expensive to do the experiments initially on that; we are doing it in a wheat variety that can be grown as a winter wheat" ("Newsnight" 2012).

According to Woodward, many conventional farmers said the trial was "irrelevant" ("Newsnight" 2012). The argument disputing the relevance of the experiment was repeated by Theo Simon on Channel 4 when he asked Graham Jellis: "Tell me, as someone involved in the R&D for the wheat crop in this country, what wheat farmers have approached you in the last ten years and said: 'we want GM wheat, we want you to develop genetically modified spring wheat with aphid resistance?' Have you had any demand for this that you use public funds?" The television journalist repeated the question: "Have farmers asked you to do this here in this country? Yes or No?" Jellis answered: "It is not quite simple as that!" (Channel 4, 2012). GM Freeze and Take the Flour Back argued that it would be quite unthinkable to argue that farmers, in general, support these field trials,. This is so because adopting GM wheat would cause serious problems for British exports. Wheat exports account for 15–20 per cent of the annual UK crop, with around two million tons exported each year. During the 2010–11 season, the UK exported almost double the export surplus, demonstrating the strong demand for UK wheat. The five top customers for British wheat are the Netherlands, Spain, the United States, Portugal and Germany (HGCA 2012). In Europe, there is virtually no market for GM food, let alone GM wheat; there is a very strict labeling, and traceability process (which we explain in the next section), and public opinion is increasingly opposed to GM food (European Commission 2010a). In the United States, GM wheat is undesirable, and Monsanto dropped its plans for such a modified crop amidst fervent commercial resistance from farmers around the world (BBC 2004). Guy Smith, a former GM trial farmer as well as a long-time member of the biotech industry lobby group CropGen, recently reiterated those viewpoints by declaring that the wheat trial is not just a massive waste of public money but little short of insane: "Take the latest Rothamsted work looking at GM insect-resistant wheat [...] the main question I have is why are we spending a large chunk of our finite R&D budget on a crop no one wants to buy? Even in the USA, GM wheat has stalled because of consumer resistance. Can anyone think of another example of money being spent on the development of a crop that has no market prospects?" (Matthews 2012). In order to further argue for the pointlessness of the study, in March 2012 GM Freeze, The Real Bread Campaign and Ethical Consumer wrote to hundreds of bakeries, mills, supermarkets and restaurant chains asking for their company policy on using GM wheat as an ingredient or handling GM wheat in their facilities. Not all of them, but certainly the vast majority, stated quite adamantly (and sometimes in a proud tone) that they did not use GM ingredients whatsoever (GM Freeze 2012b).

Who will benefit?

If GM skeptics dismissed the arguments that aphid repellent GM wheat would help decrease the use of pesticides and help increase yields, and argued that farmers were certainly not going to benefit from GM wheat (even if the experiments had proved to be successful), then how did they interpret the Rothamsted field trials?

During the Channel 4 debate, both Theo Simon and Graham Jellis agreed that these trials were not just about experimenting with GM wheat, but had a much broader significance. Simon's opening remarks during the debate were: "This isn't really a test of aphid resistance; I think this is a test as far as the biotech corporations are concerned of the European resistance. I think this is an attempt by the biotech corporations to start to soften up the climate in this country towards bringing back GM crops." The journalist then asked: "Graham Jellis, this is exactly what you want to do, isn't it? You want us to think again about this technology ten years after we kicked it out the door," Jellis replied: "Yes, we certainly want it to be thought about again. It's been a very successful technology in many parts of the world; 10 percent of the world's crops are now GM. Now, that's often forgotten when we think about the European situation" (Channel 4, 2012).

What was—and still is—lying behind the fears of some NGOs, farmers, associations and individuals about biotech corporations is the idea that the seeds would be patented, and farmers' practices would be constrained by the framework imposed by the company selling the seed. Patents play a crucial role in R&D. The process of patenting involves the granting, authorized by a government, of exclusive rights to an inventor/company/laboratory for a specific period in exchange for the artifact's public disclosure. Patents are an extremely contestable issue as it is debatable whether patenting spurs (Eisenberg 1987) or hampers (Jaffe and Lerner 2004) progress. Patents in the field of agribiotechnology were made possible in 1980 when the US Supreme Court ruled in the case of *Diamond v. Chakrabarty* that living micro-organisms were patentable. In 1985, the US Patent and Trademark Board of Appeals broadened this decision and determined that in principle, any plant could be patented. Two years later, the Patent and Trademark Office ruled that all living animals could, theoretically, be patented (Kleinman 2005, 58–59). By 1998, the US Patent and Trademark Office had issued 1,370 patents for agricultural biotechnology products. Seventy-four percent of these patents were held by just six corporations (Clapp 2007, 41). The plant-science industry is one of the world's most R&D-intensive sectors and ranks in the top four global industries in terms of percentage of sales invested in R&D. More precisely, the industry's top ten companies invest

$2.25 billion, or 7.5 percent of sales, in the research and development of cutting-edge products in crop protection, non-agricultural pest control, seeds and plant biotechnology (EuropaBio 2012d, 50–51). In the case of the GM wheat under discussion, Rothamsted stated that the ideas behind the aphid-repellent plants "have not and cannot be patented" because they had been previously discussed in various publications (Rothamsted Research 2012e). "The results will be freely available as soon as practicable after the experiments are completed. Data will be published in peer-reviewed scientific journals and disseminated via other relevant media" (ibid.). In fact, the idea that the results of the field trials would be "freely" available to the public was a recurrent motif stressed by Rothamsted in their correspondence with Take the Flour Back before and after the planned attacks endorsed by the NGO (*Guardian* 2012). While this is true, Take the Flour Back argued, bringing a GM crop to market entails such high costs that only a "biotech company with the required financial clout" would be able to do so (Take the Flour Back 2012b). Therefore, the counter-argument suggested that the fact that the GM seed had not been patented did not mean it would be freely available to farmers. The results may be freely accessible to the public, but it will take an agbiotech company to produce and eventually sell the GM seeds to farmers (GMO Safety 2012). Such concerns were inadvertently consolidated by Pickett himself, the program's principal investigator, who told *Farmers Weekly* that "companies are very interested and they are keeping a watching brief as they always do in all research […] We have all been wined and dined very heavily by academic groups in the USA wanting to see what new things we have got in mind for GM […] This is of global, great significance and it could be that we generate very good intellectual property for commercial development in the interests of the UK and European agriculture and business. That's the long-term plan, but this trial has no commercial connection whatsoever" (Case 2012b).

Distribution

Appropriation in Distribution refers to the social interactions among actors in their struggle to control various aspects related to how a GMO is distributed in a market. It is rather obvious that such interplays among social actors can occur simultaneously across different social locations and over short or long stretches of time. One may find it interesting to examine, for example, what led a particular country finally to allow the import or cultivation of certain type of GM maize, how a large supermarket chain decides to place GM products on their shelves, or how a local population chooses to boycott GMOs. While the distribution of the Rothamsted GM Wheat in the EU market is

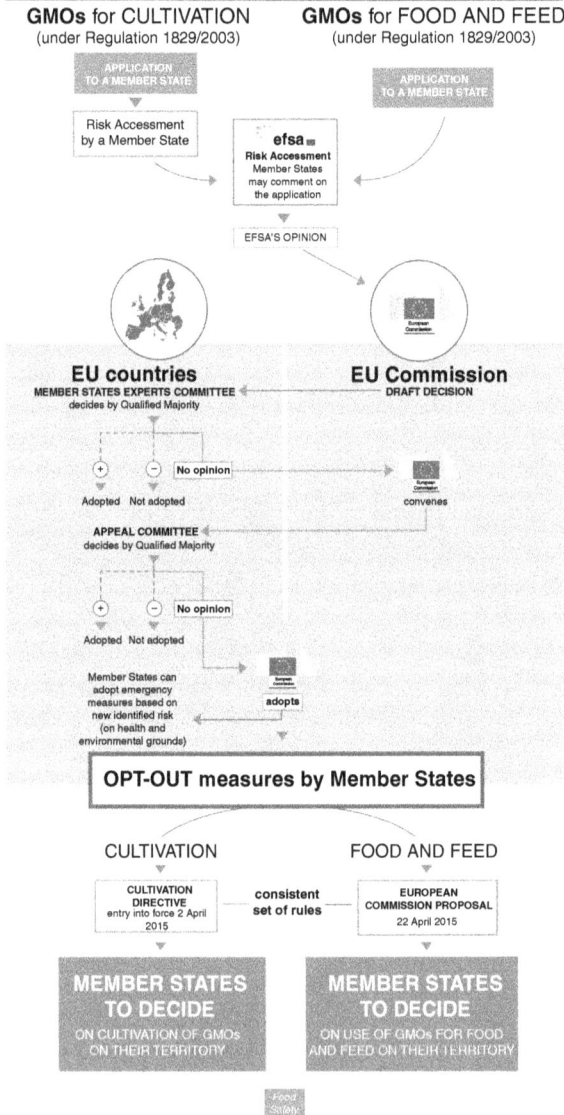

Figure 6.2 The decision-making process in the EU. (*Source:* European Commission).

now an extremely unlikely event given the disappointing results, it is still very relevant to discuss the overarching elements that make possible/legitimate the very presence of GMOs in a European Union member state. Doing so will help better anchor the Rothamsted conjuncture in the broader sociopolitical context and also facilitate the discussion to move towards different time/space locations. The focus of this section, therefore, on the authorization process in the EU.

The authorization process for commercialization

In order to better understand the nature of unfolding social interactions in the struggle for appropriation in the phase of distribution, we first have to review how decisions regarding the authorization of GMOs are reached. The approach followed in the EU concerning GMOs is a "precautionary" one "imposing a premarket authorization for any GMO to be placed on the market and a post-market environmental monitoring for any authorized GMO" (European Commission 2015a). The authorization process is carried out at the EU level, but—based on Directive (EU) 2015/412[9]—each member state is now allowed to opt out of cultivation of GMOs in their territory regardless of the European Commision's (EC) decision and the European Food Safety Agency's (EFSA) opinion.[10] Figure 6.2 is a guide to EU legislation.

Application Submission   Requests for GMOs, either for purposes of cultivation or for food and feed, can only be allowed on the market once they have received authorization. The process for authorizing a new GMO is based on the EU regulation on GM food and feed (Regulation (EC) No 1829/2003) and the EU traceability, labeling, and derived food and feed Regulations (Regulation (EC) No 1830/2003). The main objectives of these two regulations are to "protect human and animal health through stringent safety assessment of GM food and feed before it can be sold; ensure common procedures for risk assessment and authorization are efficient, transparent and do not take too long; ensure clear labelling that responds to the concerns of consumers (including farmers buying feed) and enables them to make informed choices" (European Commission 2012b).

An application for authorizing food or feed consisting of, or made from, a GMO must be submitted to the national authorities. The application must abide by the requirements of Regulation 1829/2003 and must, therefore, include:

• Studies showing that the GM food is not dangerous to health or the environment;

- Analyses showing that the GM food is *substantially equivalent* to conventional counterparts; (e.g., by analysis of particular constituents/nutrients);
- Suggestions for product labeling;
- Methods and sample material for detecting GM content;
- Indication of confidential information;
- Summary of the application dossier.

The national authority must acknowledge receipt of the application within 14 days and, in turn, send the application to EFSA for a risk assessment (GMO Compass 2006).

Safety Assessment   EFSA must then make the application summary available to the public. If the application covers cultivation, EFSA delegates the environmental risk assessment to an EU country that sends EFSA its risk-assessment report. No matter where in the EU the company applies, EFSA is responsible for assessing the risks and potential threats the GMO(s) may present to the environment, human health and animal safety. EFSA usually performs the risk assessment within six months of receiving the application and issues a scientific opinion published in the *EFSA Journal*. If, however, it decides to request more information from the applicant, risk assessment takes longer. Once the procedure has been concluded, EFSA submits its opinion to the European Commission and EU countries. The opinion is made available to the public, except for certain confidential aspects. Once EFSA publishes its risk assessment, the public has 30 days to comment on the Commission website for applications under Regulation 1829/2003, and on the Joint Research Centre website on the assessment report of the "lead" EU country for applications under Directive 2001/18.

Final Decision: Within three months of receiving EFSA's opinion, the EC prepares a draft implementing the decision, granting or refusing authorization. If the EC has a different opinion from EFSA, it should justify that in writing. The EC submits its draft for a decision to the member states, represented in the Standing Committee on the Food Chain and Animal Health. If the committee votes "Yes," the EC adopts the draft decision. If the member states, through the standing committee, vote "No" or if the result is "No Opinion"—that is, no qualified majority is expressed either in favor or against—the EC submits the draft decision to the Appeal Committee.[11] In the Appeal Committee, the member states vote a second time on the draft decision suggested by the EC. If the member states approve of the draft, the EC adopts the decision. If they vote "No," the EC rejects the draft decision. If the result of the vote is "No opinion," the Commission is required by the GMO legal framework and by the Charter of Fundamental Rights to adopt a decision on the application (European Commission 2015c). Regardless of a potential approval at the EU level, the newly adopted Directive (EU) 2015/412 gives member states the ability to prohibit or restrict the cultivation of the GM crop in their territory. In particular, member states may adopt the "opt-out" clause either during the authorization procedure—by asking that the geographical scope of the application be amended so as to ensure that its territory will not be covered by the EU authorization—or after a GMO has been authorized—on grounds related to environmental or agricultural policy objectives, town and country planning, land use, socio-economic impacts, coexistence and public policy (European Commission 2015a).

The significance of EFSA

Some methodological clarifications

It has probably become apparent by now that the authorization procedure in the EU involves a number of social actors (individual and collective), a series of legislative actions, a significant number of risk and safety assessments, a battery of tests and, in general, an overall complex bureaucratic mechanism. Now, for those who are willing to examine how different groups try to control/appropriate the authorization process *as a whole*, they are apparently presented with two fundamental but contrasting methodological choices. The first choice is to become a TPA floater, allude to some form of ontology-in-general (i.e., by referring to legislation and structures regardless of specific applications for commercialization) and sketch the appropriative canvas with very broad brush strokes. The second choice is to become a contextualizer and position the study at the ontic level by following carefully how actions related to *specific* applications (obviously already submitted and assessed by EFSA), member states, and the Commission have unfolded over time and space. I am not entirely convinced that following either of the two options can prove of significant heuristic or explanatory value. First and foremost, both options seem to be methodologically utopian in the sense that there are so many actual and virtual entities involved that, in the end, the contextualizer's account will most likely be helplessly elliptical, and the floater's explanation profoundly trivial and/or misleading. Furthermore, an account at the level of abstract ontology will probably feel as if it is detached from the previous section on Technology, where we encountered *specific* actors, "listened" to some of the exact dialogues they had and "observed" detailed events that unfolded in the Rothamsted GM wheat site. It would be very difficult to see how connections can be made between this previous assessment, with its various specificities, and a discussion that remained at a very abstract level and had very little, if anything to say, about the GM wheat. On the other hand, it is also very difficult to see how a very detailed assessment of *specific* GMOs that have entered the authorization process could help offer a general, yet still telling, picture of how appropriative forces meet in distribution. I am very much afraid that a narration staying at the ontic level of some other GMO application will be so reliant on specificities and circumstances that it will not help us focus on aspects and processes relevant to the purposes of the investigation, but will most likely get us lost in a welter of detail.

Having said this, another thing that may have become apparent from the EU authorization procedure is that EFSA plays a crucial role in the commercialization (or not) of GMOs in the EU. "EFSA is like the High Court: after EFSA gives an opinion, it becomes more difficult for a country to return to its earlier risk assessment" (qtd. in Levidow and Carr 2010, 152). The scientific opinion expressed by the Authority serves as a point of reference, or what is called a "Draft Decision," which the public and representatives of the member states may sustain or refute. Therefore, from the point of view of actors with an interest in promoting or hindering the commercialization of GMOs, EFSA constitutes part of their external environment of action; that is, stage one in the structuration cycle. What follows is that macro actors—who possess high amounts of economic, political or cultural capital and have economic, political, cultural/moral or other reasons for wanting to advance/block the distribution of GMOs—will most certainly not consider this structural enablement/constraint in a taken-for-granted manner, but will try to shape the external environment according to their goals.

As a result, it seems to me that one methodologically realistic and quite sufficiently representative way of highlighting the appropriative interactions that unfold in distribution is to examine the role of EFSA. It is interesting to highlight the criticisms raised against EFSA from two competing standpoints and examine how actors with opposing interests try to influence the agency. While insight gained from this discussion will not be generalizable in any simple sense, it can still help us discern some broad tendencies that appear to take shape in appropriation. Since EFSA's advice holds great significance in the decision-making process, appropriative attempts will be of equally rigorous effort. In other words, actors with an interest in advancing/blocking the distribution of GMOs will mobilize a considerable amount of resources to achieve their goal.

What I am arguing, therefore, is: since an endeavor to accurately capture the dynamics of the authorization process *as a whole* is methodologically utopian, focusing on EFSA may prove to be a realistic and fruitful alternative. By first placing EFSA at the center of analysis, by examining its constitutive elements, its mode of operation and its functions, and then moving on to map the social actors and their attempts to influence the Authority brings us closer to understanding appropriative actions in distribution. In doing so, I use both types of methodological bracketing; TPA floating and ACA contextualizing. The former is primarily used to analytically traverse between dispersedly situated social actors, conjunctures and social locations and the latter when positioned actors are quoted expressing their own views and orientations towards GMOs. If the idea of landing the methodological hot-air balloon atop EFSA seems legitimate, we can move on to examine this entity and the actors surrounding it by using both TPA and ACA.

The state of GMOs in the EU is in a constant flux. One month before the results of the Rothamsted field trials were announced, and as the STC inquiry was unfolding, the 2015/412 EU directive was adopted, bringing about significant changes to the EU legislative framework by giving back "full responsibility" to member states over the cultivation of GMOs on their territory. What is more, since the beginning of the field trials, EFSA has undergone internal restructuring and has also strengthened its GM risk assessment framework by publishing new guidance for the agronomic and phenotypic characterization of GM plants (EFSA Panel on Genetically Modified Organisms 2015b) and for the renewal applications of GM food and feed (EFSA Panel on Genetically Modified Organisms 2015a). Regardless of these structural changes, EFSA's opinion has uninterruptedly played a central role in the EU authorization process since the EC's draft decision is, by and large, a reaffirmation of EFSA's eisegesis.

In this section, I examine EFSA's internal organization, institutional function, and significance in the EU decision-making process by referring to the conditions of action and events that took place at the time the field trials were in progress. I will do so for two reasons. First, in order to anchor the analysis on a coherent time/space continuum that respects the corresponding structural environment and the meaning situated agents of the time attached to it. Second, doing so will help explain how actions of that time/space period contributed to the changes in EU legislation and EFSA's internal organization; which we now, as external observers, take for granted. In this sense, we can see how the outcomes of actions in the structuration cycle became new conditions of action in late 2015. Despite these changes, the main criticisms towards EFSA and the overall authorization process discussed in this section are still dominant in the broader GM controversy. The "politicization" of the debate, the "conflicts of interest" and "revolving doors" in key scientific positions and the "unduly slow" authorization process are major themes broached in this chapter and exploited more fully in Chapter 7. Finally, the occasional alteration in the use of tenses signifies the things that are still applicable at this writing and those that are not.

The birth of EFSA

EFSA was set up in January 2002 following a series of food crises in the late 1990s—such as Bovine spongiform encephalopathy (BSE)—in order to protect European consumers from food-related health risks (EFSA 2012e). This is an independent European agency, funded by the EU budget, and which operates separately from the European Commission, European Parliament and EU member states. EFSA functions as an independent source of scientific advice and communication on risks associated with the food chain. A large part of the work undertaken by EFSA is a response to specific requests for scientific advice; subsequently, the advice offered is used to inform the policies and decisions of risk managers. EFSA undertakes scientific work, either after applications for scientific assessments are received from the European Commission, the European Parliament and EU member states, or on its own initiative. EFSA's advice is mainly related to the European legislation on food or feed safety, to regulation on substances such as pesticides and food additives, or to the development of new regulatory frameworks and policies in the field of nutrition in general. While EFSA is not involved in management processes, its advice functions as an independent scientific foundation (EFSA 2012f).

EFSA, which employed approximately 450 staff members at the time of the field trials, started an internal restructuring program in 2011 and by late 2012 was organized in five directorates overseen by the Authority's executive director—namely, three science directorates (Risk Assessment and Scientific Assistance, Scientific Evaluation of Regulated Products, and Science Strategy and Coordination), the Communications and External Relations Directorate, and the Resources and Support Directorate.[12] The science directorates offer support to the work carried out by EFSA's Scientific Committee and panels. In the case of GMOs, the directorate of interest is that of Scientific Evaluation of Regulated Products (REPRO), which supported EFSA's work in the assessment of substances, products and claims intended to be used in the food chain in order to protect public, plant and animal health as well as the environment (EFSA 2013).

EFSA signifies the separation of risk assessment and risk-management tasks within the EU and is the principal risk assessor in Europe. The Authority's mission is to evaluate risks associated with the food chain by collecting and analyzing existing research and data. It does not have scientific laboratories and does not generate new scientific research. The role of EFSA is to offer scientific advice to the risk managers—that is, to the EC, the member state authorities and the European Parliament (EFSA 2014).

Risk assessment of GMOs

As with all issues pertinent to food safety, EFSA carries out the risk assessment of GMOs in the EU. EFSA's GMO Panel is responsible for preparing and adopting the GMO risk assessments while the EU member states participate throughout the risk-assessment process. Based on the GMO Panel's risk assessment, EU member states and the European Commission decide whether to approve or reject GMO authorization applications. Despite the risk assessment provided by EFSA, EU risk managers also take into account other considerations in deciding whether or not to authorize a GMO.

Each GMO application received by EFSA is subject to a review by the independent scientific experts of the GMO Panel. These experts are responsible for evaluating the scientific information, discussing its content and deciding on the final risk-assessment opinion. The panel approaches each GMO risk assessment by following the framework laid down in the corresponding EC Directive and Regulations. In GMO risk assessment, a comparative assessment is made between the GM crop plant and its conventional crop counterpart to determine if the GM plant is as safe as its conventional non-GM counterpart. During the risk assessment process each of the following elements is considered for all applications:

- The molecular characterization of the GM product;
- The compositional, nutritional and agronomic characteristics of the GM product;
- The potential toxicity and allergenicity of the GM product;
- The potential environmental impact following a deliberate release of the GM product and taking into account its intended uses either for import, processing or cultivation.

As EFSA does not carry out its own studies, it is the applicant's responsibility and obligation to demonstrate the safety of the GM product in question and present a full application before evaluation. Since the cost of these studies must be borne by the applicant with a commercial interest in obtaining approval, EFSA is not involved in the conduct of experiments and analyses. If EFSA's GMO Panel has any doubts about the data presented or needs further evidence, the applicant is asked to provide additional data or information before the Panel delivers its final risk assessment (EFSA 2012d).

Members of EFSA's panels undertake to act independently, and their independence is ensured by a mandatory Declaration of Commitment of Independence and a Declaration of Interests (DoI). These declarations are publicly available and are also made annually. In addition, prior to each meeting all experts involved are required to declare any interests which might be considered prejudicial to their independence in relation to the items on the agenda. EFSA's policy on DoIs is not to ban or sanction the holding of interests by individuals operating in the sphere of EFSA, but to ensure transparency in the handling of situations in which potential conflicts may arise (EFSA, 2012c). It is not only the members of the GMO Panel who are required to declare their interests, but also the additional experts who work in collaboration with the GMO Panel as members of a special working group (ibid.).

Nonetheless, despite EFSA's proclaimed devotion to assessing GMO applications in a transparent and timely manner that ensures human and environmental safety, the Authority has quite often received criticisms from social actors who hold contrasting positions on GMOs.

EFSA and the authorization process: Two contrasting views

In 2011, the European Association for Bioindustries (EuropaBio[13]) published a 64-page report focusing solely on the analysis of the EU authorization process for GMOs. The *Approvals of GMOs in the European Union* report was drafted on EuropaBio's "own initiative for the benefit of all parties involved in the EU authorization process: EU member states, the European Commission and the European Food Safety Authority (EFSA), and stakeholders who rely on an efficient and workable authorization system for GM products: trait developers, seed companies, grain traders, food and feed industry and farmers" (EuropaBio 2011b). In the very detailed document, eight issue areas in the authorization process were isolated, and twenty-two recommendations were suggested to make the whole process more efficient (EuropaBio 2011a, 23). All the major problems identified by EuropaBio appear to fall broadly into two categories. The first relates to issues that render the authorization process not only very drawn out but also erratic, while the second problematic area is identified in the *politicization* of the overall process.

The process as time-consuming

The EU authorization process is quite frequently criticized on the grounds that it is time-consuming and unnecessarily complicated. In the 2012 Agricultural

Biotechnology Council report, *Going for Growth* (Agricultural Biotechnology Council 2012a), the EU approval process is criticized on numerous occasions and blamed for huge economic losses for the British economy. "The malfunctioning process of EU approvals for cultivation and import of GM crops is one factor deterring private sector investment […] The UK sector must also maintain its competitive edge and prepare to be one of the first to reap the benefits of any improvement in the EU approvals process […] Public policymakers must act now to prevent further loss of intellectual capital and to unlock the huge potential inherent in the UK's science and R&D base. Otherwise, we risk not only falling further behind the US and emerging economies but also losing our advantage within Europe" (Agricultural Biotechnology Council 2012b, 8, 14, stress added).

The same line of criticism was repeated in November 2012 by Lord de Mauley, the newly appointed parliamentary undersecretary of state, when he characterized the EU approval process as "unduly slow." De Mauley also added: "We want the EU regime to operate more effectively, grounded on an objective appraisal of the potential effects of GM crops on human health and the environment" (EuropaBio 2012b). In January 2013, the UK's Farming and Environment minister, Owen Paterson, reiterated the same issues of concern by telling reporters at the Oxford Farming Conference: "We need to work with like-minded partners to move the legislation along at a European level because it is going grindingly slowly and we are getting further and further behind" (Euractiv 2013b).

Indeed, agribiotech corporations very often stress the economic benefits EU farmers miss on out due to the EU regulation and relate the adoption of biotechnology with immediate financial advantages. A study by the University of Reading has estimated that "if the areas of transgenic maize, cotton, soya, oilseed rape and sugar beet were to be grown where there is agronomic need or benefit then farmer margins would increase by between [$570 million and $1.2 billion a] year" (Park et al. 2011, 396). If just GM maize were planted across the EU, the estimated increase in annual income would be no less significant, as Table 6.2 demonstrates.

In another study, Brookes and Barfoot claim that since 1996, "farm incomes have increased by $64.7 billion" thanks to GM crops (2011, 8), and 57 percent of this profit was due to increased yields (EuropaBio 2012d, 26). Nonetheless, EU farmers are said to be unable to enjoy such high profits offered by the competitive benefits of GM crops because of the problematic authorization process (EuropaBio, 2011b, 20).

There is no doubt that the authorization process in the EU is far more complicated and time-consuming than the corresponding processes in the United States, Canada and Brazil. In fact, the average time required for a

Table 6.2 Estimated increase annual income if GM maize was planted across EU.

Country		From € M	To € M
Bulgaria		3.6	5.4
Czech Republic		4.6	9.2
Germany		25.7	42.4
Greece		1.2	5.9
France		34.2	85.5
Italy		40.6	108.2
Hungary		6.2	12.6
Austria		12.0	16.8
Poland		11.9	29.9
Portugal		1.4	2.4
Romania		12.1	21.5
Slovakia		3.6	5.9
Total		**€157 million**	**€334 million**

Source: EuropaBio (2012, 25).

GM product approval in the EU is 45 months, while for the United States, Brazil and Canada GMOs are approved after 25, 27 and 30 months respectively (EuropaBio 2011a, 11). Partly blamed for this situation is the European Commission, which is criticized for disregarding the three-month deadline set by EU legislation and for spending over 11 months before it asks member states to vote on a GM product (EuropaBio 2011a, 6). One reason for such delays is that, despite the long waiting list of products that should be voted on, voting procedures are very often canceled. More precisely, in 2010 five out of eleven standing committees for member states to vote on GM products did not take place. Delays were even greater the following year, as out of the six standing committees scheduled for the first half of 2011; five were canceled (EuropaBio 2011a, 23). Time is also wasted when standing committees adopt the practice of grouping product votes; that is, collecting three or more products to vote on only a few times per year. The practice of GM products with a positive EFSA opinion not following a chronological order in the voting process causes significant discomfort to applicant corporations, as this unpredictability has a potential impact on trade and/or cultivation scheduling (EuropaBio 2011a, 23).

The politicization of the authorization process

The second major point of discontent raised against the EU authorization process is the *politicization* of the science-based system. According to EuropaBio, one facet of this politicization is the fact that on numerous occasions member states tend to ignore EFSA's positive scientific opinion on a GMO application and vote against it on sociopolitical grounds. Countries such as Austria, Luxembourg and Greece have voted in all cases against EFSA's scientific opinion (EuropaBio 2011a, 29), while the negative trend in votes is also witnessed in countries such as Poland, Hungary, Cyprus and Malta (EuropaBio 2011a, 31). The director of Green Biotechnology Europe at EuropaBio, Carel du Marchie Sarvaas, believes that "using cultural reasons to opt out of a science-based process undermines the credibility of the system [...] Socio-economic reasons would not stand up in a court of law and there are also questions about whether those reasons are compatible with international trade law or European internal market law" (Earls 2011). This need to disentangle politics from science and speed up the process has also been expressed by Anne Glover, former CSA of the EC, who has argued: "I think we could really get somewhere in Europe if when evidence is used partially, there were an obligation on people to say why they have rejected evidence [...] The evidence with which I work is independent, the evidence with which I work does not change according to political philosophy. And that should give people a lot of confidence" (Euractiv 2012c).

The EuropaBio trade group sees as another problematic facet of politicization the fact that the European Commission has quite often intervened in EFSA's risk-management measures in ways not endorsed by EFSA's scientists. More specifically, data requirements such as the 90-day feeding studies in rats and the presence of antibiotic resistance markers are two risk-management measures that EFSA scientists have deemed unnecessary (EuropaBio 2011a, 28). These extra measures are seen as superfluous by EuropaBio, as they do not ensure higher safety, but merely render the whole process more time-consuming. Although GMO applications are examined by EFSA on a case-by-case basis, Anne Glover seems to be implying that such management measures are indeed unnecessary. A few months after her appointment in the EC, Glover stated that: "There is no substantiated case of any adverse impact on human health, animal health or environmental health, so that's pretty robust evidence, and I would be confident in saying that there is no more risk in eating GMO food than eating conventionally farmed food [...] Most of us forget that most plants are toxic, and it's only because we cook them, or the quantity that we eat them in, that makes them suitable" (Euractiv 2012c).

In order to overcome these two systemic deficiencies, EuropaBio has presented an extensive set of measures that will help speed up the process and alleviate the enormous backlog of GM products found in the EU process. Almost in their entirety, the recommendations advocated by the group of Europe's largest agribiotech corporations are centered around the role of EFSA. EuropaBio seems to show great affinity for an authorization process that would be based on the following principles:

- Member states' votes should be contingent on the EFSA opinion;
- Commission and member states should guard over EFSA independence;
- New elements of EFSA guidance should not be applied retroactively to dossiers already in the process;
- The accumulation of applications should be dealt with through a particular action plan implemented by EFSA. (EuropaBio 2011b, 7)

For GMO advocates, therefore, the authorization process is unnecessarily complicated, as they see no actual risks being involved in the commercialization of GMOs. What is more, EFSA is regarded as mildly inadequate, as its members are likely to yield to the fear of political cost and not reach decisions on "objective" grounds.

EFSA as the playground of the biotech industry: Conflicts of interest and revolving doors

GM skeptics campaign for two exact same things; namely, EFSA's independence and greater emphasis on scientific evidence. Quite naturally, however, these two key elements of the authorization process are now interpreted in a strikingly different manner.

For skeptics of GMOs, EFSA will become genuinely independent so long as it stops being nothing more than the "playground of the Biotech Industry" (Testbiotech 2012). EFSA's credibility has come under fire, as there have been cases of conflict of interest by which members of the Authority exploit their official capacities for personal or corporate benefit. While, as it was stated previously, EFSA members have to state explicitly their interests on an annual basis and before each meeting, EFSA's selection criteria do not include independence from industry (EFSA, 2012c). In February 2012, the campaign

Table 6.3 Reported conflicts of interest at EFSA, 2010–12 (2010).

When?	Who?	What?
24 March 2010	Suzy Renckens (GMO Panel)	Head of the Secretariat to the EFSA GMO Panel takes lobbyist job at Syngenta (revolving-door case). Testbiotech/Corporate Europe Observatory joint complaint.
29 September 2010	Diana Bánáti (Management Board)	EFSA management board chair Diáná Bánáti's conflict of interest case with ILSI Europe. Bánáti resigned from the board of ILSI Europe and was re-elected chair of EFSA's management board on October 21.
29 November 2010	Laura Smillie (Risk Communication Unit)	EUFIC revolving-door case.
1 December 2010	Harry Kuiper (GMO Panel)	ILSI conflict of interest case.
23 February 2011	Milan Kovac, Matthias Horst, Jiri Ruprich, Piet Vanthemsche (management board)	Conflicts of interest of four management board members with Danone, ILSI, EUFIC and COPA.
7 April 2011	Angelo Moretto, Alan Boobis, Theodorus Brock (PPR panel)	Conflicts of interest rife with Europe's pesticide and food-safety regulators.
15 June 2011	ANS Panel	Eleven out of 20 experts on panel on food additives have a conflict of interest, as defined by the OECD. Four members of the panel fail to declare active collaborations with ILSI Europe.
13 September 2011	Ursula Gundert-Remy, Riccardo Crebelli (ANS Panel)	Two of five experts newly appointed in July were found to be in violation of internal EFSA rules because they had failed to disclose consulting activities for ILSI.
27 October 2011	Albert Flynn (Chair of NDA Panel)	NDA panel chair Albert Flynn has conflict of interest related to Kraft Foods.
7 November 2011	GMO Panel	Twelve out of 21 experts on GMO Panel have conflicts of interest, as defined by the OECD.
19 December 2011	EFSA working group on TTC	Ten out of 13 members of EFSA TTC working group have a conflict of interest.

Source: Corporate Europe Observatory and Earth Open Source (2012, 24).

groups Corporate Europe Observatory and Earth Open Source copublished a report on the credibility and independence of EFSA titled *Conflict on the Menu: A Decade of Industry Influence at the European Safety Authority*. The document presented EFSA members' conflict of interest and the issue of "revolving doors"[14] as undeniable and harsh realities. Table 6.3 (Corporate Europe Observatory and Earth Open Source 2012, 24) demonstrates reports on conflicts of interest that appeared just for the years 2010–11. In fact, reports on the credibility of EFSA date back to 2004, two years after the Authority was established, when Friends of the Earth (FOE) reported on the GMO Panel (Friends of the Earth Europe 2004). Among other issues raised, FOE expressed serious doubts as to whether EFSA could provide an independent and scientifically based decision when several members of the EFSA GMO Panel had at the time direct financial links with the biotech industry, and two of them even appeared in promotional videos produced by the biotech industry (Friends of the Earth Europe 2004, 3).

Concerns over EFSA's independence have been raised not only by civil society groups, but by a number of social actors across different domains, such as members of the European Parliament, public institutions, scientists and the media (Corporate Europe Observatory and Earth Open Source 2012, 2). One compelling case of conflict of interest and revolving doors was the Bánáti incident. In 2010, José Bové, a representative of the Greens in the European Parliament (EP) and vice-president of the EP's agriculture committee, pointed to the fact that Diáná Bánáti, the chairwoman of the administrative board of EFSA, was at the same time working as a member of the governing council of the International Life Science Institute (ILSI). ILSI is an organization representing businesses that are also members of EuropaBio, such as Monsanto, Syngenta and Dupont, but also other corporations such as Nestlé, Coca-Cola and Kraft, and has a history of lobbying for the interests of the food, chemical and biotech industries (Corporate Europe Observatory and Earth Open Source 2012, 3). At the press conference Bové said:

> Given the highly sensitive nature of the European Food Safety Authority's work, notably in the EU's GM authorization process, the need for the independence of its staff is crystal clear. That the chair of the EFSA board should have such direct links to the food industry is a clear conflict of interest and completely unacceptable. Ms. Bánáti's continuation as chair of the EFSA is no longer tenable and her resignation should be tendered immediately […] The Commission should never have approved her appointment given her clear links to the food industry, which is completely at odds with the need for independence at the EFSA. However, it is equally concerning that the Commission has

failed to respond to the concerns about Ms. Bánáti's independence that I raised with the responsible Commissioner John Dalli[15] in July of this year. There can be no alternative but to replace Ms. Bánáti as chair of the EFSA. However, this scandal raises further doubts about the EFSA and its work. (The Greens in the European Parliament 2010)

Following these allegations, Bánáti resigned from her position at ILSI and was re-elected as chairwoman of the administrative board at EFSA. In 2012, however, she moved back to ILSI to become executive director and scientific director and was asked to resign from her position at EFSA[16] (EFSA 2012a).

As an aftermath of the Bánáti incident, the European Parliament decided to withhold approval for the 2010 budget of EFSA due to there being a clear evidence of conflicts of interest and revolving doors (Euractiv 2012b). In 2011, following a complaint from German NGO TestBioTech, the European Ombudsman, Nikiforos Diamandouros, "concluded that EFSA had not carried out as thorough an assessment of the alleged potential conflict of interest as it should have" and called on EFSA "to strengthen its rules and procedures to avoid potential conflicts of interest in 'revolving door' cases" (European Union 2011). The complaint concerned the case of Dr. Suzy Renckens, head of EFSA's GMO Unit, who was connected to Syngenta. Following the unfolding of such events, EU member states seem to become quite wary of the conflicts of interest at EFSA. In June 2012, a committee representing the EU's member states decided not to approve the nomination of Mella Frewen to the board of EFSA following pressure from the European Parliament. Frewen, currently the director-general of FoodDrinkEurope, had in the past worked as a lobbyist for Monsanto, and it was deemed that her current and previous job positions were incompatible with a role in EFSA. On hearing the news, she stated: "It is both unfortunate and disappointing that the whole process became so politicized on a point of principle of having an industry representative as a legitimate member of the EFSA management board, as set out in the EFSA founding regulation" (Keating 2012).

While the presence of industry representatives is indeed not excluded by EFSA's regulation, their presence, as may have become apparent by now, causes great distress to a variety of social actors. This is so because individuals related to the industry are said to strive for favoritism in EFSA's risk assessment. There are numerous cases in which the Authority has been accused of intentionally disregarding independent scientific studies that raise concerns about the safety of—not necessarily GM—products already approved by EFSA. Some examples include Monsanto's MON863 GM maize and the toxic effects found by Gilles-Eric Séralini (Séralini, Cellier and de Vendomois

2007), the health risk posed by glyphosate (Antoniou, Habib et al. 2011)—the main ingredient in the widely used GM companion herbicide Roundup invented by Moloney—and EFSA's decision to raise the limit on glyphosate residues two hundred-fold (EFSA 2012b), and the use of aspartame as a widely used artificial sweetener that has been found by some studies to be carcinogenic (Soffritti et al. 2007).

The birth of the comparative safety assessment

While these examples may, on their own, constitute very interesting cases for further analysis, it appears that what would be more telling about the impartiality of EFSA is the examination of one of the very constitutive elements of its GMO risk assessment principles: namely the notion of *Comparative Safety Assessment*. This idea serves as the "general principle for the risk assessment of GM plants" and is the overarching element in the Guidance for the Environmental Risk Assessment (ERA) issued by EFSA (EFSA 2010, 12). "The purpose of the ERA is to assess if the introduction of the GM plant into the environment would have adverse effects on human and animal health and the environment [...] The underlying assumption of the comparative assessment for GM plants is that the biology of traditionally cultivated plants from which the GM plants have been derived, and the appropriate comparator is well known" (EFSA 2010, 11). This concept helps simplify risk assessment as it avoids a more comprehensive assessment of GM plants. Numerous scientists consider this approach as insufficient as *by definition* it excludes hazards from synergistic non-linear effects[17] (Then 2010a). The concept of *comparative assessment*, as used by EFSA, is said to have been developed by industry and ILSI between 2001 and 2003 (Then 2010b, 3). In the summer of 2000, Harry Kuiper was elected as chairperson of the joint working group between the Food and Agriculture Organization of the United Nations (FAO) and the World Health Organization (WHO) on "Safety aspects of genetically modified foods of plant origin" (FAO/WHO 2000). That working group discussed the "comparative approach" for the very first time. From 2001, Kuiper worked as author of the "Task Force of the ILSI International Food Biotechnology Committee." At that time, members of the task force represented companies such as Cargill, Syngenta, Dupont/Pioneer, Bayer CropSciences, Monsanto and more (ILSI 2004). Kuiper was "one of the most influential experts in Europe" on risk assessment, publishing several papers in his field of expertise and at the same time chairing the EU project ENTRANSFOOD—a project on the risk assessment of GM plants—supported by the Commission and industry (Then and Bauer-Panskus 2010, 5). In 2003, Kuiper and Kok—both members of the ILSI task force—suggested that the notion of Comparative Assessment should be used as a starting point for testing GMOs (Kok and Kuiper 2003). That same year EFSA's GMO Panel was established, headed by Renckens—who, as we previously saw, later moved to Syngenta—and chaired by Kuiper. The team around the two experts worked on EFSA's guidance document for the risk assessment of food and feed derived from genetically modified plants (EFSA 2004). Comparative assessment was placed at the heart of the risk-assessment procedure and has remained despite EFSA's revisions of the guidance (EFSA 2011). The fact that EFSA had shared ILSI's views/recommendations on the safety assessment of GMOs was seen as a huge success by ILSI's task force. During a 2006 workshop in Athens, the chair of the ILSI task force, Kevin Glenn from Monsanto, remarked on the 2004 report:

> In 2004, the task force's work culminated in the publication of a report that included a series of recommendations for the nutritional and safety assessments of such foods and feeds. This document has gained global recognition from organizations such as the European Food Safety Agency and has been cited by Japan and Australia in 2005 in their comments to Codex Alimentarius. The substantial equivalence paradigm, called the comparative safety assessment process in the 2004 ILSI publication, is a basic principle in the document. (qtd in Then and Bauer-Panskus 2010, 7)

Instances such as those mentioned in this section make skeptics of GMOs very wary of the role of EFSA. Instead of merely rejecting this technology on sociopolitical grounds, it appears that social actors who are not in favor of GMOs would like the Authority to become disentangled from economic interests and phenomena of revolving doors, and be more open to independent studies that argue for more comprehensive tests on the health and safety issues potentially posed by GMOs.

Concluding remarks and clarifications

In this section, I have tried to offer some understanding of the ways that appropriative struggles unfold. While it has been argued from the beginning that no generalizations or contextless conclusions can be drawn from this discussion, it is hoped some insight has been gained. The evidence provided so far suggests that, in the GM debate, two main tendencies appear to take shape. On the one hand, social actors with an interest in promoting GMOs are more likely to stress the *financial* benefits derived from this technology and downplay the potential risk factors. On the other hand, those actors who try to obstruct the spread of GMOs have apparently as their primary concern the avoidance of *risks* related to health issues and/or environmental damage. The case of EFSA is very interesting. Although established as an independent and autonomous authority, it has received fierce criticism from both camps of the GM debate. GM advocates criticize it as a carrier of political interests, while GM skeptics construe it as a puppet in the hands of financial interests. It is true that the authorization process has quite often been inexcusably delayed; especially in cases in which the European Commission itself has violated the three-month voting deadline set out in EU legislation. It is equally true that the issue of "revolving doors" and "conflicts of interest," no matter how circumstantial, cannot be taken in a light-hearted manner.

By examining EFSA, I hope to have brought to the foreground the tensions between profit and a more disinterested scientific governance that takes risk seriously. In light of this, it is argued that claims for an impartial or disinterested science within policymaking should not be taken at face value, but should be very carefully interrogated. It has also been suggested that by situating processes and decisions within sets of structural positions and relations, potential financial, political or cultural interests can be revealed. Finally, it is worth stressing that revealing influences stemming from various institutional spheres, and how these shape science within policymaking, is an *analytically separate* question from the competing stances on a scientific issue per se. Some insight into scientific debates on GMOs was offered in Technology, as were the concerns raised by Steinbrecher regarding the scientific team at Rothamsted. These competing stances were expressed using purely the tools of science, and any financial, political or cultural interests were certainly excluded from the discussion. We can now turn to discuss central Ideological elements encountered, not only during the debate on the field trials, but also that routinely recur throughout the broader controversy.

Chapter 7

THE ROTHAMSTED GM WHEAT
TRIALS (II): IDEOLOGY

Ideology

The third and final facet of TAI—Technology, Appropriation, Ideology—
refers to the discursive constructs that may prompt social actors to support
or oppose the research, production and/or distribution of GMOs. As mentioned in Chapter 4, ideology can be witnessed in the discourse of broad and
abstract interpretative nature, but can also be discerned in notions that *appear*
to be precise referential statements. It would very interesting to explore the
connotations and significance attached to two concepts couched in the two
extremes of the ideological spectrum: one notion seemingly being a technical
term, and the other being a "mundane" word of everyday usage. By doing so,
a very rich and vivid undercurrent of interests, beliefs, (convenient) interpretations, position-practices and more may be revealed flowing under discursive
elements that either *appear* to be technical terms or are used in a *supposedly*
consensual frame.

In this chapter, I focus on two notions that appear to play a central role in
the GMO debate. The first refers to the ways the concept of *nature* is construed,
while the second refers to the notion of the precautionary principle. These two
concepts appear in the GMO debate in very different ways. While the concept of the term *nature* is *by definition* involved in the discussion of GMOs as
the technology itself is all about modifying the genetic material of organisms
that exist in nature, the significance attached to it is quite often *implied* and the
various connotations of nature need further elaboration in order to be fully
understood. On the other hand, while the precautionary principle is part and
parcel of the GM debate, it is a highly contentious concept that often appears
in discussions *as if* it is a scientific issue while, in reality, it is an ideological construct subject to contrasting interpretations.

It is hoped that by the end of this chapter the importance and intercon-
nectedness of these two notions will have become clear. They are both key
moments in the discursive orientation to GM, and the stance that one takes
regarding the "nature of *Nature*" has a significant effect on the attitude to the

precautionary principle. As the exploration of these two notions unfolds, some of the actors, decisions and position-practices discussed in the previous chapter make their appearances once again, though under a different light.

The nature of Nature

As already seen in the previous chapter, on the day of the planned "decontamination" of the GM wheat trials, Maurice Moloney stood in front of the field site and defended the objective of the experiment by suggesting that Rothamsted researchers were arguably "the greenest people on the planet" (Channel 4, 2012). This deeply ideological statement immediately begs the question as to who and what is Green; and if it is indeed within the sincere intentions of the Rothamsted researchers to safeguard nature and the environment, why are environmental groups opposing them? Is there a paradox, then, that a shared environmental outlook instigates protest? Or is there a wider divide in paradigms lying concealed behind common ideological discourse?

In order to assess how nature is construed in the GM debate, I first quote pertinent beliefs and actions of the various social actors who have appeared throughout Chapter 6 and, subsequently, examine whether these stem from the same or competing strands of thought. I start with collective and individual actors who have expressed their skepticism or outright opposition to GMOs:

- One of the chief aims of GeneWatch is to "ensure that genetic technologies are developed and used in the public interest and in a way which *protects human health* and *the environment* and *respects human rights and the interests of animals*" (GeneWatch 2013, stress added);
- Friends of the Earth declare they stand for "a *positive relationship* with the environment" and the planet on which "we depend [and should] *keep in good shape*" (Friends of the Earth 2013, stress added);
- The individuals and organizations brought together under the umbrella of the GM Freeze campaign are "united in seeing an immediate need to stop and consider the massive social, economic and *environmental effects* GM is having worldwide" (GM Freeze 2013, stress added);
- EcoNexus, whose report to DEFRA against the environmental release of the Rothamsted GM Wheat was quite extensively analyzed in the Technology section, "investigates, reports and acts on *threats to biodiversity, climate, ecosystems, local and agro-ecological farming systems, food security & sovereignty*, health and the interests of indigenous peoples and local communities" (EcoNexus 2013, stress added);
- The Take the Flour Back network called their plan to uproot the Rothamsted GM Wheat an act of "decontamination" (Take the Flour Back

2012), signifying, therefore, that agribiotech products introduce something unwanted or even toxic to the natural environment.

Apart from the "programmatic declarations" of collective actors regarding nature, individual actors have also expressed their views in equally clear terms. Greenpeace chief Lord Melchett, who was arrested for attacking a GM crop trial farm at Lyng, Norfolk, in 1999, said: "I was very pleased that we managed to remove some of the *genetic pollution* [...] The chemical industry think they have some sort of private right to plant this stuff and cause this *pollution to the environment* and this threat" (BBC 1999, stress added). José Bové has very often stressed the "need for environmental protection" (Bové 2001, 92); Antje Lorch from EcoNexus called the Commission's authorization for cultivation of BASF's GM potato Amflora as an "authorization for contamination" (Lorch 2010); Christoph Then from Testbiotech has been actively involved in the campaigns No Patent on Life! and No Patents on Seeds, and is a special adviser for Greenpeace (Testbiotech 2013); and the codirector of EcoNexus, Helena Paul, has among other activities edited the books *Healthy Crops, A New Agricultural Revolution*, which argues against the use of chemicals and for natural processes that can strengthen crop resistance, and *The Forest Within*, which is an anthropological study of how the Tukano Amazonian Indians construed their "partnership" with the rainforest (EcoNexus 2010).

Environmentalism and GM skeptics

The list of such examples could be much lengthier, but even from this rather short representation of instances it becomes apparent that, at the very least, what skeptics or opponents of GMOs share is an affinity towards nature; a belief that nature should be safeguarded and protected from the side effects of human activity. In other words, one can discern a sort of environmentalist arc being formed by GMO skeptics. For Robyn Eckersley (1992), there are five major currents of environmentalist thought and action: *resource conservation, human welfare ecology, preservationism, animal rights* and *ecocentrism*. The *resource conservation* perspective proceeds from a human-centered and utilitarian framework that seeks "the greatest good for the greatest number" by reducing waste and inefficiency in the consumption and exploitation of nonrenewable natural resources (e.g., natural gas, oil and more) and ensuring a maximum sustainable yield of renewable resources (e.g., crops, timber, soil and so on). Although this perspective is inextricably linked to the production process and construes the nonhuman world in use-value terms, it can still be regarded as the first, but major, stop away from an unrestrained development approach (Eckersley 1992, 35–36). *Human welfare ecology* breaks away from the

"narrow, economistic focus of resource conservationists" by being primarily concerned with the overall state of health and resilience of the physical
and social environment (Eckersley 1992, 37). While still anthropocentric, this
strand of environmentalism focuses on the physical and *social* limits to growth
by stressing the health, psychological and recreational needs of human communities (ibid.). *Preservationism* entails a sense of aesthetic and spiritual appre-
ciation of wilderness, which could be described as reverence (Eckersley 1992,
38). In this current of environmentalism "priority is given to untouched 'wild
nature', or wilderness, to be respected, even regarded as sacred, independently
of any instrumental human interest in it" (Benton 2007, 83). *Animal liberation*,
which has developed relatively independently from the previous streams of
environmentalism, champions the "moral worthiness" of certain members of
the nonhuman world. This advocacy of animal rights clearly constitutes a
form of demarcation from purely anthropocentric facets of environmentalism (Eckersley 1992, 42). Finally, *ecocentrism* builds on the principles of preservationism by espousing the aesthetic and spiritual considerations raised,
but becomes a more wide-ranging approach by expressing the need to "protect threatened populations, species, habitats and ecosystems *wherever situated*
and irrespective of their use-value or importance to humans. (This kind of
concern is well illustrated by the activities of the international environmental organization Greenpeace)" (Eckersley 1992, 45). The ecocentric approach
"favors choices that minimize the harm to nonhuman nature consistently with
maintaining human flourishing" (Benton 2007, 83).

It should be quite obvious that I am not arguing that the beliefs held by
skeptics of GMOs are homogeneous. In fact, within Green thought there can
be instances of conflict as proponents of *resource conservation* constitute "somewhat of a foe to more radical streams of environmentalism" (Eckersley 1992,
35). What does seem to be the case, however, is that in the very least actors
who are not willing to readily espouse GM technology share an environmental vision that in its most modest form argues that human development and
progress should be restrained—even if this stance is simply a reflection of our
own self-interest. The differences that José Bové and Helena Paul may have
regarding the issues of animal rights or overpopulation could be significant
from an environmentalist point of view, but would be rather negligible if one
were to reject environmentalism in its entirety and adopt a contrasting strand
of thought.

Technological determinism and GM advocates

At first glance, advocates of GMOs seem to share some of the environmental sensitivities that GM skeptics/opponents express. Sustaining biodiversity,

reducing greenhouse gas emissions and ensuring the safety of food are issues that appear to be high on the agenda of Rothamsted Research (Rothamsted Research 2012g, 3–8), BBSRC (2009, 6–8), EuropaBio (2012d, 41–42, 45, 47) and the Agricultural Biotechnology Council (2012a, 4, 7). One could even argue that such principles dovetail with the aspirations of *resource conservation*. What is more, in the same line of discursive argumentation, advocates of GMOs portray agribiotechnology as an entirely natural process. This claim is made in two different senses. First, agribiotechnology is portrayed as a *continuation* of previous agricultural practices of selective breeding and hybridism: that is, as "a process of innovation which has been going on for thousands of years and which is continued by scientists around the world today" (Agricultural Biotechnology Council 2012a, 2). Second, agribiotechnology is depicted as a practice that involves processes occurring in nature independent of human intervention. This "naturalness" of agribiotechnology has been explicitly stressed by Professor Moloney, who has explained that "gene transfer" is a "natural process" by arguing: "Gene transfer between species that do not normally undergo sexual crossing sounds unnatural to the layperson. However, the development of gene transfer techniques is mainly based on copying systems already present in nature that have been moving genes around for hundreds of millions of years" (Moloney 1995). In fact, the claim that biotechnology is a *natural* science, utterly compatible with nature, has constituted a central pillar of Monsanto's promotional campaign and has helped sustain the equation "biotechnology = nature = safe" (Kleinman and Kloppenburg 1991, 433–34). In this mode of thinking, therefore, nature is more or less represented as a qualitative yardstick against which agribiotechnology is verified as a beneficial and harmless form of science and technology.

While in the first instance the images and discursive elements related to nature serve as a vehicle for introducing agribiotechnology as a benign form of human intervention, what usually follows is an implicit portrayal of nature as inherently weak or inadequate to tackle the challenges that arise from human activity.[1] Very often agribiotech companies attempt to promote GMOs as a response to the mounting environmental problems. Such problems include the steady decline of the ratio of arable land to population, climate change (including an increase in water scarcity and droughts), an increase of carbon emissions and air pollution and more (EuropaBio 2012c, 39, 41–42; CropLife International 2013). These challenges can be met with the cultivation of GMOs, which are endorsed for their alleged higher yields, their decreased use of insecticides and pesticides, soil preservation, reduction of CO_2 emissions and much more[2] (EuropaBio 2010a,b, 2012d, 2013c; ISAAA 2013). GMOs are, therefore, offered as an answer to some of the most pressing global challenges and as a way to fulfill the "need to improve plants," as the European biotech industry puts it (EuropaBio 2012d, 10).

It gradually starts to become apparent that the environmental concerns of agribiotech corporations are based on entirely different grounds than those expressed by GMO skeptics/opponents. This is so because, at the very least, the primary tenet of the environmentalist arc argues for a moderation/ restriction of human activity that is harmful to the environment. On the other hand, GMO companies present human activity and its accompanied environmental problems in a taken-for-granted manner and pose no questions as to whether new orientations to the environment are required. The orienting frame of agribiotech corporations is one of alleviating unwanted consequences[3] through technological solutions that compensate for anything lost. For these social actors, the answer to the stress human activity puts on nature is found by further encouraging human activity; a kind of human endeavor that "improves plants" and, therefore, in some sense "fortifies" nature against (human-made) environmental threats. In other words, agribiotech companies are heralding a future in which science and technology will help humanity protect itself from environmental/natural threats and put an end to insecurity. "Biotechnology is then just the latest and most extreme example of the desire to control and subjugate nature" (Lee 2008, 35). This consequently raises the question: If GMO skeptics espouse an environmentalist mode of thinking, what do GMO advocates embrace?

One can rather safely place the ideological advocacy of agribiotech companies within the arc of technological determinism.[4] Following the work of Benton (1994, 2007) on ecology and the nature/society dualism, it becomes clear that GMO advocates appear to construe nature from the prism of human mastery.

> In the most influential version of the technological determinist view—sometimes referred to as "cornucopian," or "Promethean," or as "technological optimist"—there are two aspects of mastery of nature. One is that we should become increasingly able to protect ourselves from formerly catastrophic threats from nature: storms, floods, droughts, diseases, predators, and so on. The other aspect is that nature should become an indefinitely expanding reservoir for the satisfaction of human desire. Science and technology promise an end to poverty, insecurity, and disease, and a prospect of ever-growing material prosperity and cultural enrichment. (Benton 1994, 32)

These two facets of the technological determinist view are exactly the two main messages that agribiotech companies endeavor to get across in the "mission statements" discussed previously. GMOs can and should be, the argument goes, used to moderate the side effects of human activity.

Like the environmentalist strand of thought, here too there are variations within the technological determinist mode of thinking such as the Western capitalist cornucopian version, the state-socialist variant, and the "managerialist" approach (Benton 1994, 31–38). The cornucopian model in the industrialized capitalist societies purports "good life in terms of ever-expanding individual choice in consumer goods which satisfy the full range of human needs and desires" (Benton 1994, 32). The state–socialist variant that prevailed in "formerly actually existing" state–socialist societies de-emphasized individual choice for the sake of distributive justice and the destruction of capitalist private property (ibid.). Third, the managerialist approach, pioneered by the Club of Rome, "wears an eco-friendly mask" and argues that there are objective, nature-given limits to human growth and expansion. Although this approach calls for control and regulation of human activity—if necessary—it is still couched in the basic thesis of technological determinism, as it cannot offer any solutions outside technological innovation and "the already dominant paradigm of growth and development" (Benton 1994, 37). Finally, regardless of whether a social actor espouses any of the three forms of technological determinism, it becomes obvious that the endorsement of GMOs as an answer to the threatening environmental problems highlighted by agribiotech companies and institutes constitutes a "technological fix" solution. For Benton, a "technical or technological fix" is a corollary of technological determinism (Benton 1994, 37) and refers to the "tendency to rely on technological innovation to solve social and ecological problems" (2007, 81).

Concluding remarks

In conclusion, what appears to be the case in the GMO debate is that advocates and skeptics of this kind of technology seem to be aligned within two different ideological camps as to how nature is construed. Although there are variations within these two major strands of thought, skeptics of GMOs abide by an environmentalist outlook on nature, while advocates of GM crops espouse some form of technological determinism (or optimism). Although some common discursive elements such as "biodiversity" and "reduction of CO_2 emissions" are shared by both groups, it soon becomes apparent that such terms have entirely different meanings for the two groups. GMO advocates take for granted human activity and the need to gratify the model of the "good life" and suggest that the side effects will wither away via "technological fixes." On the other hand, environmentalists argue, even in their mildest form, a need to moderate human activity so as to live a harmonious life alongside nature, on which we depend. It is hoped that it has become apparent that while on the surface there seems to be congruence between the two groups regarding the "nature of *Nature*," with closer examination deep rifts between skeptics and advocates clearly emerge. Another point of serious dispute is the concept of the precautionary principle, now to be discussed.

The precautionary principle: What's in a name?

A widely established—and intensely debated—response to the challenges of risk, uncertainty and ignorance in innovation policy is the precautionary principle. Despite being a constitutive element of international risk-management protocols, it remains a subject of much misunderstanding and mischief (Stirling 2014, 59). As already seen from the framing and evidence submitted to the STC Inquiry in Chapter 1 and the arguments expressed by the former EU CSA in Chapter 6, the principle quite often comes under fire for purportedly opposing scientific reason, delaying the authorization process and stifling technoscientific innovation in general. In this section, I clarify what the precautionary principle is about and assess whether these accusations are couched in an accurate understanding of the term.

The precautionary principle is a bedrock principle of the most influential protocols and regulations on environmental protection and human health. Its application is unequivocally stated in the 1992 Rio Declaration on Environment and Development (UNESCO 1992, 3); it is reaffirmed in the Cartagena Protocol on Biosafety (Secretariat of the Convention on Biological Diversity 2000, 2); it is a constitutive part of the Treaty on the Functioning of the European Union (European Union 2008), and is an integral component of the 2001/18 EU Directive on the Deliberate Release into the Environment of GMOs (European Commission 2001, 1) and EC Regulation 1830/2003 concerning the traceability and labeling of GMOS (European Commission 2003c, 1). In these documents, the precautionary approach is presented in a way that suggests that scientists and policymakers are fully aware of the principle's definition and connotations[5].

The fact that in these risk-management frameworks, the precautionary principle is presented as an undisputed approach that helps safeguard human, animal and environmental safety constitutes some sort of paradox, as no clear definition of the precautionary principle exists. The irony is that "where the precautionary principle is unobjectionable, it is simply common sense; where the principle goes beyond common sense, it is objectionable both in theory and in practice" (Weale 2007, 592). The reason for this potential confusion is that "the precautionary principle has neither a commonly agreed definition nor a set of criteria to guide its implementation" (O'Riordan and Jordan 1995, 3).

Despite constituting "an important principle of EU law, and hence also of the regulation of GMOs, the precautionary principle is subject to vastly different interpretations and roles" (Lee 2008, 43). Therefore, "in many ways we should not be surprised that there is controversy over exactly what claims are being made when the precautionary principle is asserted as a guide to policy" (Weale 2007, 592).

While the precautionary principle is apparently a relatively open-ended concept that can accommodate a rather wide range of approaches, there are some distinctive characteristics that help narrow down its definition and implications. From an ideal-typical perspective, there appears to be a consensus (Kriebel et al. 2004, 146) that—at the very least—the principle is couched on four central components:

(1) Taking preventive action in the face of uncertainty;
(2) Shifting the burden of proof to the proponents of an activity;
(3) Exploring a broad range of alternatives to potentially harmful activities;
(4) Increasing public participation in decision-making (Raffensperger and Tickner 1999, 350).

Despite the fact that the precautionary principle rests on four major tenets, it has received criticism for a much greater number of reasons (P. Saunders 2010, 48–49). In this chapter, I assess two of the four constitutive elements of the principle: namely, the ideas of *preventive action* and *democratic decision-making structures*. The former is explored in the following section as it will prove to entail important "details" of the precautionary principle that certainly need to be spelled out, while the latter is discussed in the ensuing chapter, where the GM controversy is appreciated in relation to the broader social order. While the precautionary principle has received an astonishing amount of attention, and there has been a huge amount of scholarly work on this issue (Ellis 2006, 446), I seek to be as succinct and comprehensive as possible in order to demonstrate what lies behind the principle and clarify what is at stake if this approach is eventually abandoned.

Risk avoidance: Between Utopia and reality in a globalized world

The precautionary principle acknowledges the place of scientific uncertainties at the center of decision-making in the sense that "scientific uncertainty does not in itself preclude regulatory action" (Lee 2008, 42–43). In simpler terms, at the core of the precautionary principle lies the idea that "decision–makers should act in advance of scientific certainty to protect the environment [so that] risk avoidance becomes an established-decision norm where there is reasonable uncertainty regarding possible environmental damage or social deprivation arising out of a proposed course of action" (O'Riordan and Jordan 1995, 3). Therefore, "its focus is on currently unidentified but potentially significant, and possibly irreversible, future effects of introducing particular technologies when there is a lack of theory and evidence with which to make regulatory decisions" (Welsh and Ervin 2006, 155). "The precautionary principle recommends a policy that guards against risk where there is uncertainty about how, whether and to what degree those risks will materialize" (Weale 2007, 593). Such regulatory action should be contingent on a cost–benefit analysis and the severity or irreversibility of the (potential) risk (UNESCO 1992, 3). In other words, there should be careful "examination of identifiable social and environmental gains arising from a course of action that justifies the costs"[6] (O'Riordan and Jordan 1995, 4).

The notion of "risk avoidance" is an extremely contestable issue. The very idea that potential environmental risks should be avoided opens up the discussion of what *kinds* of risks should be avoided/accepted and the *extent* to which such risks should be avoided/accepted. How is one to decide the ramifications of "potentially *significant* and possibly *irreversible*" effects of a new technology and/or the acceptable threshold of a "*possible* environmental damage?" There is, therefore, a qualitative side (what kinds of risks) and a quantitative side (how likely, and to what extent, can these risks unfold) of the precautionary principle which make its interpretation a rather subjective issue.

Consequently, it comes as no surprise that different social actors uphold varying versions of the precautionary principle. One example is Greenpeace, which stands for the "zero risk" version of the principle. For this NGO, a "zero-risk"–"zero-tolerance" policy should be adopted: "If a polluter cannot prove that what he is discharging will not damage the environment then he simply isn't allowed to discard that sort of waste" (Weale 2007, 592). Friends of the Earth appear to adopt a similar version of the principle by stating that: "Action should be taken to prevent harm to the environment and human health, even if scientific proof is inconclusive" (Amis 2010); in other words, Friends of the Earth dismiss "harm" in general without leaving room for a quantitative and qualitative framing of the issue. On the other hand, the British government—which first formally accepted the principle in its 1990 environmental strategy, published as *This Common Inheritance*—advocates a much milder version of the principle:

> Where there are *significant* risks of damage to the environment, the Government will be prepared to take precautionary action to limit the use of potentially dangerous materials or the spread of potentially dangerous pollutants, even where scientific knowledge is not

conclusive, if the balance of likely costs and benefits justifies it. (House
of Commons: Environment Committee 1990, 11, stress added)

Therefore, the relative vagueness that surrounds the notion of "risk avoidance" opens up the possibility of milder and stronger forms of the precautionary principle to coexist. What follows from this obvious realization is that
one cannot express a viewpoint on the principle *as if* it means one and only
one thing. Consequently, Anne Glover's adamant stance against the need to
adhere to the precautionary principle during the decision-making process
(Euractiv 2012d) is at best too ambivalent a viewpoint. To what kind of principle was the EU's CSA referring when she said it was "no longer relevant"
(Euractiv 2012c)? Was she suggesting that risks should *not* be avoided? Should
all new technologies be implemented without any prior scientific evaluation?
Glover's stance bypasses the milder/stronger form distinction by apparently
rejecting the notion in its entirety. But, how is it possible that the EU's former
CSA was against the principle, and the executive director of Green technology at EuropaBio, Nathalie Moll, clearly stated that "the precautionary principle is good and it should be used and it is used in Europe" (Euractiv 2013a)?
Could it be that the former implicitly rejected a "zero-risk" version while the
latter supported a milder form of risk avoidance? This could certainly be the
case, as the formal position of EuropaBio on the precautionary principle was
developed in a 1999 discussion paper and appears to respect the main tenets
of the approach (EuropaBio 1999). More specifically, it is stated:

> EuropaBio believes it is imperative that there be a common under-
> standing and a consistent interpretation of the term "Precautionary
> Principle." An unclear definition could lead to demands for "zero risk"
> which is unachievable and would also have serious consequences on inno-
> vation […] Moreover, the European biotechnology industry would like
> to stress that as a responsible industry, it abides by legislation embodying
> a precautionary approach in its requirements for risk assessment and
> a precautionary principle in defining risk management measures prior
> to market release (e.g., Directive 90/220/EEC). (EuropaBio 1999, 1–2)

Consequently, in some cases, while the principle is in general accepted, its
"zero-risk" version is rejected and more often than not becomes subject to
harsh criticism. The most common objection to this extreme form of the precautionary principle is that it stifles any form of innovation and progress. This
is so because it is tough to argue for a new kind of technology the application
of which entails absolutely zero risk. In fact, at a meeting held at the Royal
Institution in 2003, 40 scientists argued that if the precautionary principle

were applied in its "zero-risk" form the following innovations would not have materialized: heart surgery, antibiotics, airplanes, bicycles, high-voltage power grids, pasteurization, pesticides and biotechnology (Tudge 2003). Another similar problem that renders this rigid version of the principle a utopian project is that technologies currently in use are not without any risk to humans, animals and the environment. Therefore, since pesticides currently in use harm the environment in some way, on what grounds should new technologies with the same or even smaller degrees of environmental damage be banned? What seems to be the problem with this account of the precautionary principle is that it ignores the risks already inherent in the status quo and, consequently, renders the demand that all innovations be *absolutely safe*, impossible in practice and utopian in principle (Weale 2007, 598).

To repeat the argument developed so far: while one of the constitutive elements of the precautionary principle is the notion of "risk avoidance," the fact that under the umbrella of "risks and damages" lies a whole spectrum of quantitatively (extent) and qualitatively (gravity) different hazards to human, animal and environmental well-being makes it very difficult to set an objectively defined threshold beyond which the principle ought to be applied. There are, however, certain social actors such as Greenpeace and Friends of the Earth who set this threshold to "zero" by arguing that no new technology should be introduced unless its proponent demonstrates that there is no risk entailed in its implementation. On the other hand, all the United Nations member states that ratified the Rio Declaration on Environment and Development have agreed to a precautionary principle that accepts some sort of risk so long as the threats from the new technologies do not pose "serious or irreversible damage." It is true that the rigid 'zero-risk' version of the principle is utopian and ignores the hazards already present in the existing state of affairs. Nonetheless, the argument that the precautionary principle *in general* represses progress and development is misleading, as this is *only* applicable to the extreme version of the approach.

Preventive actions and irreversible damages

Having said this, there appears to be a key term in the Rio Declaration (reaffirmed in the Cartagena Protocol and the EU Directives) that deserves attention, as it seems to be pertinent to the discussion on GMOs; namely, the notion of *irreversible damage*. The idea of irreversible damage may prove to be some wedge between GMOs and, at the very least, the other technological advances that the Royal Institution scientists mentioned. It is true that innovations such as airplanes, antibiotics, heart surgeries and so on are part of reality on virtually a global scale. It is also true that there is always a risk or hazard linked with these technologies; an airplane may crash, an antibiotic may cause severe side effects, a heart surgery may prove fatal and so forth. Nevertheless, no matter how serious these risks may be, they do not have the character of irreversibility. These innovations can be contained and are always subject to reevaluation and amelioration; Boeing Dreamliners have recently been grounded until all risk suspicions are alleviated (Ostrower and Pasztor 2013), antibiotics as growth promoters have been banned (European Union 2005), and the overall death rate in coronary artery surgery declined by 21 per cent between 2001–09 in the UK (BBC 2009). On the other hand, growing concerns have been voiced about the extreme difficulty in implementing such bans, improvements or ameliorations to GMOs once these have been released to the environment.

It is not only GM canola that can cross-pollinate with non-GM canola and in effect render the latter into a crop with GM material, as we saw in the previous chapter, but it appears that this is the case with *virtually all GM crops*. The possibility of transgene flow from GM crops to their wild relatives has been recognized by numerous scientists. In fact, among the first to express such concerns were two Calgene scientists who explicitly stated: "The sexual transfer of genes to weedy species to create a more persistent weed is probably the greatest environmental risk of planting a new variety of crop species" (Goodman and Newell 1985). Crop-to-wild gene flow can create the problem of hybridization. Under the appropriate conditions, hybridization between a common species and a rare one can "send the rare species to extinction in a few generations." This has been the case in some wild subspecies of rice and also in the increased extinction risk of wild taxon (Ellstrand 2001, 1544). The escape of transgenes into wild populations can also lead to increased weediness, invasion of new habitats by the wild population and also can negatively affect native species with which the wild plant interacts (including herbivores, pathogens and other plant species in the community) (Pilson and Prendeville 2004, 165). "*Removing or recalling genes once they have escaped into natural gene pools is impossible.* There are no adequate safeguards against gene flow between the GMO and native organisms where transgenes are likely to affect fitness, decrease genetic diversity, or increase toxicity" (Altieri 2005, 369, stress added). Furthermore, while empirical evidence of long-term GM seed persistence in conventional agriculture is scarce, recent research brought to light remarkable findings. Ten years after a trial of GM herbicide-tolerant oilseed rape carried out in Sweden by Plant Genetic Systems, it was found that despite very strict seed ("volunteer") control over the years, GM seeds still survived and were subsequently successfully grown in the laboratory for the purposes of the research (Hertefeldt, Jørgensen and Pettersson 2008).

This inevitably raises the question of "coexistence"; that is, whether GM and non-GM crops can exist side by side in nature. While the European

Commission endorses legislation "on the coexistence of genetically modified crops with conventional and organic farming" on the grounds that this is "directly related to the practical choice of consumers and agricultural producers to respect individual preferences and economic opportunities" (European Commission, 2003b), and research from the Institute for Prospective Technological Studies advises that coexistence for maize, sugar beets and cotton is feasible so long as the suggested isolation distances are respected (Messean et al. 2006, 15), there are others who still argue that in practice coexistence is unattainable (Antoniou et al. 2012, 87; Altieri 2005; Ponti 2005; Müller 2003). The feasibility of coexistence is contingent on specific precautionary measures that GM farmers have to adhere to in order to ensure that nearby non-GM fields are not affected by genetically modified material. In the unfortunate event that such cross-pollination takes place, the damage is irreversible. "*In case of damage, synthetic transgenes cannot be removed— they can be expected to persist for several thousands of years in the wild plant populations. In this case, safety and ecological soundness can no longer be guaranteed.* Experience from the field of eco-toxicology of chemicals has shown that particularly persistent substances, regardless of a first assessment of their risks, have a very high ecological damage potential" (Müller 2003, 36, stress added).

Unfortunately, such concerns are not limited to seeds, but also to animals that have been genetically modified. One such example is the GM salmon developed by the Canadian biotech firm AquaBounty, which was awaiting approval by the US Food and Drug Administration (FDA) as the Rothamsted Trials (FDA 2013) unfolded. At that time, the FDA issued for public comment a draft environmental assessment related to the agency's review of the application concerning the GM salmon. FDA's preliminary finding is that approving the commercialization of GM Salmon will not pose "a significant impact on the U.S. environment" (FDA 2013). On November 19, 2015, "After an exhaustive and rigorous scientific review, FDA [...] arrived at the decision that AquAdvantage salmon is as safe to eat as any non-genetically engineered (GE) Atlantic salmon, and also as nutritious" (FDA 2015).

The AquAdvantage Salmon (AAS) includes a gene from the Chinook salmon and has been genetically modified so as to grow to market size in half the time of conventional salmon (i.e., 18 months instead of 36). AAS will be grown as sterile, all-female populations in land-based facilities. As a result, AquAdvantage Salmon cannot escape or reproduce in the wild and poses no threat to wild salmon populations. Apart from being a safe genetically modified fish, AquaBounty argues that there are economic and environmental benefits to the adoption of such a technological innovation, as the use of land-based facilities reduces the environmental impact on coastal areas, eliminates the threat of disease transfer from farms to wild fish and grows more fish with less feed. In addition, facilities located near major consumer markets reduce the environmental impact associated with air and ocean freight (AquaBounty Technologies 2013). While AquaBounty is reassuring that AAS will be grown in land-based facilities, researchers examined what would happen in the unlikely event that this type of GM salmon found its way to streams and rivers. In the experiment, it was observed that AAS could mate with wild brown trout and the new hybrid species grew significantly faster than AAS. The study concludes:

> Despite the apparent low probability for genetic introgression into the brown trout genome, the ecological consequences of decreased salmon growth in the presence of transgenic hybrids indicate that hybridization is relevant to risk assessments. Although transgenic hybrids would probably be rarer in the wild than in our experiment, our results indicate that transgenic hybrids have a competitive advantage over salmon in at least some semi-natural conditions. Still, it is entirely unclear whether this would be observed in truly wild environments. *If this advantage is maintained in the wild, transgenic hybrids could detrimentally affect wild salmon populations. Ultimately, we suggest that hybridization of transgenic fishes with closely related species represents potential ecological risks for wild populations and a possible route for introgression of a transgene, however low the likelihood, into a new species in nature.* (Oke et al. 2013, 6, stress added)

To conclude, in this section I have discussed how there is a variety of types of precautionary principle ranging from mild to zero-risk versions. I have suggested that while a zero-risk version is utopian, the mildest forms of the principle can prove to be inefficient in safeguarding human health and the environment. In order to avoid the dilemma between a utopian vision and low risk-avoidance standards, I have suggested that the notion of irreversible damage can provide more specific ethical guidance *within* the concept of the precautionary principle. In this way, the compromises that should be made with new technologies are narrowed down, and more backbone is provided to the milder version.

It seems to me that so long as one is willing to respect the emphasis that the Rio Declaration, the Cartagena Protocol, and EU legislation place on the issue of risk avoidance, it is unthinkable to argue for GMO risk management without the presence of the precautionary principle. In the event that policymakers yield to some pressures asking to circumvent the precautionary principle, then the likelihood of facing irreversible damages increases alarmingly.

There is a very significant amount of research that demonstrates that GMOs *in principle* have the potential to cause *irreversible damage* to ecosystems.

I have mentioned some of these studies in order to highlight that the danger of irreversible damage exists in theory and can very well become a reality so long as precautionary measures are not seriously followed through. Finally, I have argued that attributing to the precautionary principle the idea that it necessarily resists development and progress is profoundly misleading. I now turn to discuss the notion of public involvement that the precautionary principle foresees.

The precautionary principle: Backward and forward

At the European level, the separation of risk assessment and risk management is fundamental and enshrined by law (EFSA 2014). While EFSA is the principal risk assessor in the EU, the precautionary principle plays an integral role in risk-management directives. Despite its global recognition as an overarching policy principle (Harremoës et al. 2001), it is very often misunderstood to be a decision-making rule, while, in essence, it is a principle to inform processes of social deliberation about uncertain technologies that pose threats of serious or irreversible harm (Nightingale 2014, para. 2). The fact that the principle has a *normative* character—by alluding to processes that expend effort in social learning and exploring a broad range of salient knowledge—makes it vulnerable to unjust criticism (Stirling 2013, 250–51). More precisely, as seen in Chapter 6, the principle is quite often juxtaposed with the "science-based" regulatory risk assessment and found to be responsible for the unduly slow authorization process and heavily "politicized" EU legislation framework, which seriously compromises the European prospects for technological progress.

However, what has been suggested so far—and needs to be spelled out clearly once again—is that both risk management and risk assessment are informed by normative contents, negotiated framings and the larger historical and sociocultural forces within which they are embedded (Felt and Wynne 2007, 34–35; Stirling 2014, 58). Risk assessors perform their analyses by bringing to the table their own deeply seated convictions, value judgments and selective sensitivity to a wide range of issues (Stirling 2013, 249). The binary framing of the GM controversy, therefore, as "science vs. politics" or "progress vs. irrationality" is both artificial and misleading. The precautionary principle does not offer an alibi to scientific procrastination, political abstention and a priori rejection of a technoscientific trajectory. On the contrary, as a principle of social deliberation, it suggests a *wide range* of reasonable options should be considered and not just a single preferred subset, which may lead to cognitive inbreeding and closing down of options (Nightingale 2014, para. 15). What is more, the precautionary principle underscores the need for a healthy humility regarding the sufficiency of scientific knowledge, especially under conditions of uncertainty (i.e., do not know the odds; may identify the main parameters) and ignorance (i.e., do not know what we do not know) (Harremoës et al. 2001, 170; Wynne 1992, 114). This in effect blemishes the usually unchallenged conviction in the probabilistic analysis of risk assessment, especially in cases where an innovation pathway, such as GM technology, is novel or complex, and uncertainties cannot be reduced to single definite probabilities (Stirling 2014, 58). By accepting uncertainty and ignorance as integral parts of scientific knowledge, precaution guards against the error of considering the absence of evidence of harm as proof of absence of harm (Stirling 2014, 60). The "precautionary principle has nothing to do with anti-science and everything to do with the rejection of reductionist, closed and arbitrarily narrow science in favor of sounder, more rigorous and more robust science" (Harremoës et al. 2001, 185).

Taking the above points into account, it appears, therefore, that the precautionary principle as a policy principle accommodates most of the aspirations and concerns expressed by interested parties during the Rothamsted trials, but also across the various initiatives of public dialogue taken by the UK government throughout these years. Nonetheless, as was obvious in the evidence submitted to the STC Inquiry discussed in Chapter 1, formal political bodies as well as distinguished scientists call for its immediate repeal as a constitutive principle in EU legislation on the grounds that it undermines scientific enterprise and stifles progress. Whether these criticisms spring from genuine beliefs, vested interests or intentional misinterpretations is difficult to discern. Nonetheless, the very existence of dispute on such fundamental policy issues is very telling about the deeply schismatic nature of the GM controversy. This realization naturally leads to broader questions extending well beyond the Rothamsted trials: What is the actual magnitude of the GM debate, and what is at stake in the controversy? What does this tell us in terms of practical policy, and what can be done to alleviate the tension? I turn to these questions in the next chapter.

Conclusion

In Chapters 6 and 7, the Rothamsted GM wheat trials were mapped within the normative and institutional arrangements of the field of agribiotechnology. In order to examine the significance of the field trials in an analytically coherent manner, the research question was first framed in terms of its Technological, Appropriative and Ideological dimensions. Inside Technology, we followed the premises of SST in order to acquire a deeper understanding of how and why social and sociotechnical relations follow specific trajectories. So, for example, we explored the external environment of action that scientists at Rothamsted

Research face; how the institute is organized in terms of position-practices, who is at the helm, what funds are available, who offers these funds and under which terms, what agenda has been set for the following years and more. Apart from the conditions of action, we have also tried to get a better understanding of certain key members' dispositions. Moloney's and Pickett's deeply seated schemata of action have been cautiously inferred from the convictions expressed in their writings and their involvement in other social situations. In Technology, we also discussed details of the genetically modified wheat. By taking a closer look at sociotechnical relations, we examined how scientists at Rothamsted describe the GM wheat's natural tendencies, needs and distinctive mode of interaction with nature (i.e., how it is designed to repel aphids) and how other scientists challenge the validity of the conjuncturally specific knowledge of the former by offering a contrasting account about the safety and effectiveness of this technology. Through the examination of productive forces, both types of methodological bracketing were used: TPA when the discussion centered on factual information or, in instances where I specifically expressed my own assumptions or interpretation of things, and ACA when situated actors were directly quoted expressing their own beliefs and argumentations. Throughout the discussion of Technology, the loci of social space and time shifted. The hot-air balloon traveled from inside the walls of Rothamsted to the offices of EcoNexus and from there to many years earlier when John Pickett got involved in the controversial Pusztai affair. All this traveling was aimed at gaining better insight into the structural arrangements and social actors involved in the productive forces.

In Appropriation, the focus of attention moved away from the details of Technology to the ways in which various groups try to control it. In R&D and Farming,[7] we referred to the various individual and collective actors who opposed the field trials in one way or another and the corresponding replies they received from various proponents of the Rothamsted experiment. At times, quite high levels of contextualization were pursued by quoting how specific dialogues unfolded on British television so as to offer the reader a clearer sense of the tone of the debate and the commitment that some actors demonstrated to their cause. In "distribution," that is the second sub-section of Appropriation, the scope had to become broader, as the authorization process for commercialization takes places at a supranational level. The effort to offer an understanding of the main position-practices and structural arrangements involved in the process inevitably led to larger gaps in contiguity. This was clearly acknowledged when the discussion evolved around the appropriative games played for the influence of EFSA. Nonetheless, despite the gaps that methodological floating caused, there were still numerous instances when ACA was applied. We saw how think-tanks like Corporate Europe Observatory approach the issue of "revolving doors," what José Bové had to say about the case of Diáná Bánáti, EFSA's official response and more. On the other hand, we also referred to specific financial benefits the commercialization of GMOs in the EU will allegedly offer, the problems the abnormal prolongation of the voting deadline causes and the overall argument against the politicization of EFSA's science-based system. In doing so, we attempted to sketch some general trajectories that seem to emerge in the field of agribiotechnology and highlight some ways various social groups try to interpret scientific and technological advances from the perspective of policymaking.

In the second part of the empirical section, two key ideological constructs of the GM debate—the nature of *Nature* and the precautionary principle— were discussed at length. During the textual analysis of these two discursive elements, we came to see the contrasting connotations and varying interpretations these two notions have acquired within the context of GMOs. While at surface value most situated actors proclaim their commitment to *Nature* and the precautionary principle, with closer examination it soon becomes apparent that there are stark differences in the way the two terms are construed. A whole array of different emotions and suggested courses of action seem to spring from these discursive elements. More precisely, ideology favoring GM technology weaves a pattern of euphoria, of inviting humans to indulge their endless desires for a better quality of life and placing their hopes in the promises of agbiotech. It is suggested, moreover, that this is reliant on technological progress not being delayed by needless risk assessments and public involvement. Conversely, ideology opposing the flourish of GM crops invites humans to curb their desires for material well-being and urges them to try and sustain nature in its current form and adopt ways of living that endorse a more harmonious relationship with the natural environment. At the same time, this strand of ideology argues for a more open-ended policymaking procedure, one that invites the public to express its concerns and aspirations, but also more seriously takes potential risks into account. In addition, it was suggested that the viewpoints one keeps regarding *Nature* and the precautionary principle seem to be directly related. It was stressed, nonetheless, that the apparent ideological rift in the GM debate does not constitute an ironcast divide, but rather indicative tendencies of the situated actors' ideological orientation.

I now turn to this book's final substantive chapter, where the theoretical framework is used to explain larger issues germane to the GM debate, and which stretches over longer periods of historical time and space. The significance of the framework and this study to practical policy is explored by placing the GM conjuncture within the broader historical and sociocultural forces.

Chapter 8

WHAT IS THE GM CONTROVERSY? SCIENCE, POLITICS AND PROSPECTS

In the two previous chapters, the particularities of the Rothamsted GM wheat trials were examined along their technological, appropriative and ideological dimensions. Apart from issues referring directly to the field trials, broader pertinent topics of the wider GM controversy were also broached. These broader issues are now further explored in a critical discussion of the GM debate regarding its current status, distinct characteristics and prospects.

The GM Debate as a Case of Science vs. Politics?

Not only in the popular media, but also among academic, political and economic circles, the GM debate is often portrayed as a case of science against politics; that is, as an instance where scientific evidence about a technological breakthrough is facing opposition from value-based considerations and nonscientific criteria (Ammann and Kuntz 2016; BASF 2014; James Hutton Institute 2014; Kinchy 2012; National Farmers' Union 2014; Späth 2015; Syngenta 2014; *The Wall Street Journal* 2010). Within this binary interpretative scheme, the dominant processes of *politicization* and *scientization* are identified as unacceptable practices and are respectively anathematized by GM advocates and skeptics alike. On the one hand, the politicization of the regulatory system is condemned as unwarranted political interventionism in a procedure that should otherwise be based purely on scientific evidence; a recurring criticism we came across during the discussion of the EU authorization process and the application of the precautionary principle. The scientization of the debate on (ag)biotech, on the other hand, refers to the tendency to separate a controversy from its social context and transform it into a closed debate among scientific experts. This is done on the assumption that science and scientists are "the best possible arbiters of technological controversies because they are assumed to produce objective, value-neutral assessments that do not favor one social group over another" (Kinchy 2012, 2). As such, the scientization of the debate, in essence, demands that politicians ratify expert opinion and citizens to remain uninvolved during the decision-making process.

For the GM controversy to rightfully qualify as a case of scientific innovation *contra* political decision-making, the following three assumptions must hold true. First, the political field should exhibit policies that predominantly deter innovation. Second, there should be consensus within the scientific community on the safety and benefits of GMOs. Third, scientists should remain uninvolved with ideological beliefs and vested interests.

Before turning to examine the validity of these three statements, it has to be acknowledged that all of them are anchored at a macro level of ontological abstraction; that is, they refer to abstract entities and relations that exist beyond specific time/space locations. The generalizing nature of these three suggested theses is unavoidable, as they are components of an ideological construct which, as discussed in Chapter 5, is by definition abstract. Nonetheless, the validity of this construct needs to be addressed, from not only a purely theoretical/abstract standpoint but also from a situated ontology that informs the researcher about the conjuncture's specific institutional arrangements and situated agents. In the following sections, I examine the role of the political field, the existence of scientific consensus and the impartiality of scientists from both the macro and the micro scales of ontological abstraction. In this way, I approach these major components of the GM debate not only from a theoretical perspective but also from an empirical one by providing concrete examples of the EU and UK realities.

The role of the political field

The argument raised against the political field in the EU for the case of GM technology is that, by and large, it has devised such a slow and costly regulatory framework that agbiotech innovation is, in essence, stifled. As already discussed in Chapter 6, what causes the most frustration among GM advocates are the severe delays in the authorization process, which usually takes four years (House of Commons Science and Technology Committee 2014b, Q306), but can also reach 13 years—as in the case of the BASF GM Amflora potato (BASF 2014, para. 4). Taking into account the slow and unreliable regulatory system in the EU together with the high costs associated with the discovery, development and authorization of a new GM crop, estimated at $138 million, it follows that certain types of GM inventions are, in essence, blocked as they become too risky and costly investments (Syngenta 2014, para. 4).

The deteriorating health of the agbiotech R&D sector in the EU can be discerned in the shrinking number of field trials. In the UK and the EU, more precisely, the number of field trials has fallen so low that the trials are now considered "meaningless in terms of the development of agri-technology solutions" (James Hutton Institute 2014, para. 1). More precisely, the number of environmental releases of GMOs for experimental reasons in the EU started with just four in 1991, reached a peak in 1997 with 264 field trials, and decreased in 2012 to 51 field trials (European Commission 2012a). What this tells us is that political decision-making—either intentionally or unintentionally—has indeed made the UK and the EU an uninviting place for agbiotech science. But, does this mean that political interventionism hampers technoscientific progress?

In this section, I argue that this is a misinterpretation of the situation for two reasons. First, it is a lopsided understanding of social reality, as the relations between the political and scientific fields are not conflictual, but take a variety of mutually beneficial forms. Second, and following from the previous argument, policies and decisions that oppose the commercialization of GMOs—and are often identified by GM advocates as "anti-scientific" interventions—place equal importance on the findings of the scientific community. The difference, it will be suggested, lies in the fact that these political bodies espouse opposing scientific evidence—or contrasting readings of the same findings—and hierarchize differently the needs and priorities expressed in other institutional fields. Since the purpose of this section is not to provide a detailed analysis of how politics affects science and vice versa, but to assess whether the criticism of political interventionism in the case of GMOs holds true, I limit the discussion in the unidirectional way that the political institutional order and actors influence and appropriate the scientific field.

Science and politics: System integration and social integration

To appreciate in a balanced manner the diverse ways that the political field appropriates science, we should examine not only how polity and science institutionally interact as parts of the social system, but also the kinds of relationships that are instantiated among actors of the two fields. In other words, we can approach the task at hand and build a corresponding typology of relations, based on David Lockwood's fundamental analytical distinction of *system integration* and *social integration* (Lockwood 1992, 399–412).

Regarding *system integration*, we can argue that, depending on the way and the degree to which the political and the scientific fields intermesh, political influence may take three main forms. First, the nature of the political influence may be *facilitating*; that is, it may offer structural opportunities for technoscientific innovation to be materialized (e.g., attractive environment for R&D experiments) or it may adopt a legislative framework that expedites the commercialization of particular technoscientific artifacts (e.g., optional labeling and traceability of GM foods). Second, political influence can be *co-productive*.

In instances of co-production, public funds are allocated to research projects, institutional arrangements are made that strengthen ties between the private and public sector regarding R&D, public dialogue initiatives are taken and so forth. Third, the institutional political impact may be of a *regulatory/priority-setting* nature; that is, it may devise regulatory frameworks that: (a) forbid certain types of innovation (e.g., human cloning); (b) set strict guidelines as to how certain procedures shall be followed (e.g., release of GMOs in the environment); (c) set policy priorities that the technoscientific field is invited to practically acknowledge).

From the perspective of *social integration*, we can examine the types of relationships that unfold among actors from each field and the role that they play in technoscientific facilitation, co-production and regulation. What is of particular pertinence to our discussion is how political actors use technoscientific evidence in cases such as GMOs where, as we shall see in the following section, there is scientific disagreement. Contrary to the broad accusation of showing aversion to the fruits of the technoscientific field, political actors tend to utilize them as resources. For Laurent Renevier and Mark Henderson (2002), political actors often construe scientific data as a "*political artifact*" that they can exploit in their negotiations and policy-shaping argumentations (115–16). The typology developed by Connie Ozawa (1996) helps further explore the concept of "science as resource." For Ozawa (1996), science serves two major purposes in the hands of decision-makers: it functions as a "*shield*" and as a "*tool for persuasion*." Science is used as a shield when decision-makers present scientific evidence as definitive on policy decisions so as to absolve them of responsibility and from the criticisms of unhappy constituents. In this case, science is presented as arbitrating between multiple policy viewpoints or decision alternatives (Ozawa, "Science in Environmental Conflicts" 1996, 224). Once science is ascertained as a source of authority for justifying decisions, it can then be used a means of persuading the polity of the legitimacy of one policy or decision over others. In this role, science can be used either to support advocated standpoints or to preclude policy forged around an opposing scientific conclusion (Ozawa, "Science in Environmental Conflicts" 1996, 224–25). The presence of a scientific controversy does not, however, pertain only to decision-makers, but also to stakeholders with competing interests. When scientists disagree, "contending interests frequently seek to manipulate scientific advice to provide a rationale for the decision they prefer" (Ozawa and Susskind 1985, 25). In this way, scientific evidence becomes a resource utilized in cost–benefit analysis where stakeholders or decision-makers evaluate the various decisions and policies available and communicate their choices.

Before moving on to specific examples that substantiate this typology, a few remarks need to be made. First, the suggested categories are neither exhaustive nor mutually exclusive, as they may very well overlap and be further subdivided. For instance, a co-productive relation may also entail strict regulatory provisions. Second, such links between the two fields are neither predetermined nor instantiated *en bloc*. Instead, they have to be empirically examined, as they may vary both across and within networks and time/space locations. For instance, certain political bodies may use scientific evidence *x* as a tool for justifying policy *y* while the same evidence may be framed by another body in such a way so as to justify policy *z*. Third, the examination of such ties can be performed at the macro, meso and micro levels of ontological scale and through analysis, both from the standpoint of the theorist and from the hermeneutic frames of situated actors (what was discussed in Chapter 5 as TPA and ACA respectively).

System integration: Institutional facilitation, co-production and regulation

Instances of *facilitating* and *collaborative* initiatives from the political field have already been extensively discussed. We have seen, for example, from a TPA and macro ontological scale perspective, how the UK government, as part of its *Agri-Tech Strategy*, institutionalized paths of cooperation with the agbiotech industry. We have also discussed the government's numerous nationwide dialogue efforts aimed at building bridges between public opinion and agbiotech innovation. Also, we have witnessed how political bodies, such as BIS and the BBSRC, have authorized investments in GM technology, including not only the Rothamsted GM wheat trials but also other cutting-edge projects in synthetic biology and genome editing (BBSRC 2015). Further enabling and collaborative gestures were noticed in the *Going for Growth* initiative and in the government's pledge to adopt "a more cohesive science framework" across departments, to implement "improvements in the regulatory framework"[1] and to forge an overall "clear strategy for biotech" (Agricultural Biotechnology Council 2012).

A common criticism of the EC is that it exerts "too much political influence" on EFSA by intervening in the Authority's risk-assessment procedures (House of Commons Science and Technology Committee 2014b, Q314). During the STC Inquiry, Joe Perry, chair of the EFSA GMO Panel and researcher at Rothamsted for about thirty years, shed more light on the ways that "politics" of the EC intermesh with EFSA's practices by highlighting three major cases. First, Perry recognized that there can be some tensions between "politics" and "science," in cases in which the panel expresses uncertainty about specific aspects of a risk assessment and the EC, on the other hand, as risk manager, prefers a "clear and unambiguous" final text (House of Commons Science and Technology Committee 2014b, Q313). Apart from wanting EFSA to downplay the existence of scientific uncertainty so as to communicate scientific evidence to stakeholders with greater clarity, the EC has also suggested that the GMO panel should not assess the full extent of effects that herbicide-resistant GM crops may have on the environment. The reason behind this change in policy is that the EC believes the indirect effects of herbicide-tolerant GM crops on biodiversity are better assessed under pesticide regulation. This proposal has been met with skepticism, not only by the GMO Panel but also by various member states and conservation organizations (House of Commons Science and Technology Committee 2014b, Q313). A third highly publicized divergence of opinions between the EC and the EFSA GMO Panel has been the tentative inclusion of 90-day feeding studies in rodents as part of the risk assessment. In this case, the GMO Panel has suggested that these studies should not be mandatory, as they feel that "sufficient data [exist] without those" (House of Commons Science and Technology Committee 2014b, Q314, Q315). The EC, however, has ignored the advice of the panel and has included them as compulsory components of risk assessment under Regulation 503/2013 (European Commission 2013). While this incident is often construed as an example of politicization of the risk assessment procedure, with the GMO Panel accusing the EC of using "political considerations" and stating "very clearly that there [is] no need for [feeding studies] on scientific grounds" (House of Commons Science and Technology Committee 2014b, Q314), I do not think that this is an accurate reading of the situation. As we see in the next section, the significance of feeding studies has been defended by a number of scientists who have found alarming effects on the health of rodents as a result of GM food consumption. By tentatively including feeding studies in risk assessment, the EC has, in essence, responded to concerns expressed by some clusters of the scientific community with a view of revising this policy by June 30, 2016, "on the basis of new scientific information" (European Commission 2013, L157/7).

Social integration: Science as a political resource

The use of scientific evidence as a resource aimed at justifying the policies and decisions of political bodies can be noticed in numerous instances. The ill-fated Food Standards Agency (FSA) public dialogue discussed in Chapter 1 is an example of such an instance. It appears from Wynne's account, co-chair of the Steering Group, that the FSA had already decided to follow a pro-GM policy using scientific evidence as a shield against criticisms. The unwillingness of the FSA and the government to listen to alternative readings of the situation was one of the reasons that led the vice-chair of the Steering Group to resign. This became obvious when Jeff Rooker, chair of the FSA, framed public opinion as "anti-science," and Wynne was soon after informed that "policy officials and Ministers [would] consider [public input] as they please, at their own convenience, with no accountable justification or explanations or hearing of public and other stakeholder issues and concerns" (Wynne 2010a). In the resignation letter, he described the government's initiative for public dialogue as an empty gesture. "This GM-obsessed, narrow commitment from the very top of FSA only tells me that whatever the efforts we have made, and whatever we tell a dialogue-contractor, the signals from upstairs remain narrow and dogmatically narrow-minded" (ibid.).

Criticisms that the government had already planned its course of action regarding GM policy were also raised from the very beginning of the STC Inquiry. STEPS challenged the "leading" and "partial" way that the questions were formulated as a "risk" to "scientific and democratic legitimacy" (STEPS Centre 2014). The selective attitude toward scientific findings was further demonstrated in how the Inquiry's conclusions were articulated. While submitted oral and written evidence that challenged the pro-GM stance were indeed acknowledged in the STC report to the government, these were never seriously taken into account. For a series of central issues to the GM debate, such as safety of GM crops, environmental concerns, potential benefits, issues of corporate control and so forth, the evidence that supported a pro-GM stance was tacitly recognized as what seems to be the "correct scientific perspective" (Nightingale 2014, para. 17), while the opposing evidence was simply bracketed on no apparent axiological basis.

The government's policy of safeguarding technoscientific progress from "value-based" considerations can be noticed in its correspondence with the STC as well. To be more precise, in the *Conclusions and Recommendations* section of the Inquiry report, the STC urged the government and the FSA to review their public communications on GM technology and related topics so as "to ensure that these are *framed in a way that encourages constructive public debate* (House of Commons Science and Technology Committee 2015a, 70, stress added). This recommendation was suggested as a counter-action to the government's tendency to ignore the important role that value-based considerations play—along with scientific evidence—in EU decision-making about new technologies and its repeated "failure […] to lead the debate about emerging issues in science and technology (House of Commons Science and Technology Committee 2015a, 4). The STC's call for initiatives that would encourage such a "constructive public debate" was, however, dismissed by the UK government on the grounds that "science-based decision-making is the right approach on GM" and that "it would [not] be appropriate to require an assessment of non-safety factors as part of the formal decision-making process" (House of Commons Science and Technology Committee 2015b, 2). For the UK government, therefore, favorable scientific evidence on GMOs is sufficient to justify the exclusion of the public from the decision-making process.

At the same time, a different reading of the scientific evidence is offered by governments that have opted-out of GM cultivation in their lands. The member states who voted against GMOs use science, either as a shield or as a tool for persuasion, to justify the exact opposite stance from that of the UK government. In these cases, the demonstrated scientific evidence is agreeable with each member state's policy priorities which, more often than not, appear to enlist public opinion as sound components of the decision-making rationale. In describing how votes are often justified during the authorization of GMOs, Dorothée André, head of the Biotechnology Unit of the Directorate-General for Health and Consumers of the EC, notes: "A lot abstain or vote against. We do not have a lot who vote in favor [...] At the end, we hear a lot of member states saying, 'My citizens do not want it. The science is not good. I don't need this. There is no benefit'. That is the reality" (House of Commons Science and Technology Committee 2014b, Q.375). The Scottish government, for instance, appealed to "long-standing concerns about GM crops—concerns that are shared by other European countries and consumers" to justify its decision to prohibit the cultivation of GMOs (Lochhead qtd. in *The Scottish Government* 2015). The environment minister of Northern Ireland also acknowledged GM crops as "controversial" and could "potentially damage" the clean and green image of the country (Durkan qtd. in *Northern Ireland Executive* 2015), while the same motif was repeated by the Welsh Deputy Minister for Farming and Food, Rebecca Evans (Sarich 2015). Before the opt-out clause was in place, the German government's decision to ban Monsanto's highly controversial MON810 maize was also based on scientific factors. Despite its approval by the EU for commercial use throughout the bloc, Germany decided to follow France, Austria, Hungary, Greece and Luxembourg in their opposition to the cultivation of the GM maize on the grounds that "there is a justifiable reason to believe that genetically modified maize of the type MON 810 presents a danger to the environment" (Aigner qtd. in Euractiv 2009).

It has been argued so far that GM technology is not compromised by political decisions. A wealth of examples drawn from initiatives of the UK government and the EC demonstrate that the political field may intermesh with the technoscientific enterprise in a variety of constructive ways that actively encourage rather than stifle progress. Whenever the intervention is of a regulatory/priority-setting character, this is done within the duties of the political field to hierarchize the needs found in different institutional spheres and forge a corresponding policy. Some politicians and political bodies may interpret similar situations differently, while others may hierarchize priorities differently and demonstrate greater empathy for public concerns and "value-based" opinions. By arguing that different stakeholders and decision-makers interpret scientific findings in ways that better serve their interests, we have already implicitly challenged the idea that there is scientific unanimity on the safety and benefits of GMOs. I now turn to examine the stance of the scientific community more rigorously.

The stance of the scientific community on the safety and benefits of GMOs

As we have already seen, in 2015 the UK government decided to allow cultivation of GMOs on English soil after considering the scientific evidence presented to it—such as the reports of the Royal Society, the CSA, the STC, the EU Directorate-General for Research and Innovation and of other prominent scientific associations—which asserted that GMOs pose no greater risk to human health and the environment than their conventional counterparts (Council for Science and Technology 2014; Directorate-General for Research and Innovation 2010; House of Commons Science and Technology Committee 2015a; The Royal Society 2009). What has also been the case, however, is that concerns about the safety of GMOs have been raised by scientists and professional associations, alike. That was noticed, for example, in the communication between EcoNexus and DEFRA regarding the issues of transformation-induced mutations, the presence of antibiotic resistance marker genes and an overall insufficient risk-assessment deliberation of the Rothamsted GM wheat. These three issues reappeared in the broader discussion of GMOs alongside the concerns of coexistence and irreversible damage that GMOs may cause to ecosystems. But does this really mean there is scientific disagreement on the safety of GMOs? Or are these merely isolated cases of partisan scholarship? Do pro-GM scientists hold the one and only "correct scientific perspective," or are there other equally valid views? Is the scientific community divided between those who claim that GMOs are safe and those that support the opposite, or is there greater differentiation?

An artificial consensus

In 2013, more than 300 independent researchers developed and signed a joint statement—published in 2015 in the peer-reviewed journal *Environmental Sciences Europe*—which challenged the claims of a consensus on the safety of GMOs (Hilbeck et al. 2015). By evaluating the available scientific evidence published up to that point, the authors concluded that the purported scientific consensus is an "artificial construct" that is "misleading and misrepresents, or outright ignores, the currently available scientific evidence and the broad diversity of scientific opinions among scientists on this issue." While the joint statement "does not assert that GMOs are unsafe or safe," it suggests that since "no blanket statement about the safety of all GMOs is possible," these must be assessed on a "case-by-case" basis (Hilbeck et al. 2015, 1–2). As a result, instead of adopting an a priori confident or dismissive stance on the safety of GMOs, the joint statement "endorse[s] the need for further independent scientific inquiry and informed public discussion on GM product safety" (Hilbeck et al. 2015, 2).

By examining the scientific literature through the systematic reviews of animal-feeding experiments and the findings of professional societies on the health assessment of GMOS, Sheldon Krimsky (2015) reaffirmed the lack of scientific consensus remarked upon in the above joint statement. For Krimsky, the scientific literature on the health effects of GM crops fall into three main clusters. The first group of authors (cluster 1) asserts that there is no need for testing GMO products, as long as the proteins coded by the transferred genes and the host organisms are known (Krimsky 2015, 884). The second group of researchers (cluster 2) holds the exactly opposite stance by suggesting that "each GMO product must be tested for a variety of possible effects" as "science cannot, a priori, claim that a product of genetic modification is safe without undertaking a testing program that includes multiyear and multigenerational tests in animals fed on the transgenic crop" (ibid.). Finally, the third group of authors (cluster 3) asserts in their published articles that certain GMO crops that have been fed to animals "have exhibited harmful effects" compared to their conventional counterparts and, consequently, "these results should draw attention to human health concerns" (ibid.).

What follows from this fruitful typology is that the evidence presented by scientists and scientific bodies—such as the Royal Society, the STC, the CSA, the CST, Rothamsted Research, the BBSRC and more—which have argued for the inherent safety of GMOs and have influenced UK policy, do not represent the whole of the scientific community, but only a fragment of it; what now constitutes cluster 1. Conversely, two well-known cluster 2 findings are the joint statement of the 300 independent scientists, previously discussed, which claims that GMOs must be assessed on a "case-by-case" basis, and the more recent report released by the GenØk Centre for Biosafety and commissioned by the Norwegian Environment Agency, which stresses "the literature's findings on the existence of an array of possible harmful impacts" and recognizes that the "current state of knowledge on the sustainability of herbicide-tolerant crops is incomplete" (Catacora-Vargas 2014, iii). Finally, the study of Arpad Pusztai, discussed in Chapter 6, together with the other highly publicized[2] work of Gilles-Eric Séralini (2012)[3] are examples of cluster 3 research, which caution against the probable adverse effects GMOs may have on human health. It has to be clarified that, despite noticing GM-induced health effects in their animal-feeding studies, cluster 3 researchers do not hold an a priori negative stance on GMOs. Rather, they propose that "agricultural edible GMOs and formulated pesticides be evaluated very carefully by long term studies to measure their potential toxic effects" (Séralini et al. 2012, 4230).

First-generation GMOs: A case of downplayed risks and overstated benefits?

Scientific disagreement is, however, not limited to issues of GM safety, but extends to the purported benefits of GMOs as well. The global market of GMOs is dominated by a small number of products (mainly cotton, maize, soy and rapeseed) that belong to what is called the "first generation" of GMOs. First-generation products typically demonstrate two trait types: insect resistance (GM crops that have the *Bacillus thuringiensis* (Bt) toxin inserted directly into their genome, which helps confer resistance to a wide range of crop pests and ostensibly reduces the need for repeated insecticide spraying) and herbicide tolerance[4] (GM crops that have herbicide resistance genes inserted into their genome, making the crop impervious to herbicidal spraying). These types of crops, as already seen, are heralded as "green" technology which promises, among other things, increased yields and reduced pesticide use (EuropaBio 2012d, 21; House of Commons Science and Technology Committee 2015a, 11).

Putting aside the potential risks that insect-resistant and herbicide-tolerant GM crops carry—such as a threat to non-target insects, including the monarch butterfly (The Royal Society 2009, 23), evolution of herbicide-resistant "superweeds" (Nandula et al. 2005) and also the alarming research findings that link glyphosate herbicides with non-Hodgkin lymphoma (Eriksson et al. 2008) and Parkinson's disease in humans (Axelrad, Howard and McLean 2003)—the benefits of such crops are seriously doubted by the scientific literature. The STC, for example, which in its correspondence with the UK government, advises that "scientific evidence, as measured by peer-reviewed scientific publications, suggests that first-generation products have been effective in increasing crop yield and reducing pesticide use" (House of Commons Science and Technology Committee 2015a, 11), is offering only a partial reading of the scientific evidence. In university-based trials, GM soybeans have been found to give *lower* yields than their non-GM counterparts (Elmore et al. 2001) and the same pattern has been measured in the yields of *Bt* maize (Ma and Subedi 2005) and GM canola (Bennett 2009). The fact that GMOs do not offer any yield gain was also asserted by the US Department of Agriculture (USDA) in both its 2002 and 2014 reports, which stated:

Over the first 15 years of commercial use, [genetically engineered] GE seeds have not been shown to increase yield potentials of the varieties. In fact, the yields of herbicide-tolerant [HT] or insect-resistant seeds may be occasionally lower than the yields of conventional varieties if the varieties used to carry the HT or Bt genes are not the highest yielding cultivars, as in the earlier years of adoption. (Fernandez-Cornejo et al. 2014, 13)

Scientific research not only questions the claim that GM crops increase yields, but also casts doubt as to whether the main reason these crops were genetically modified in the first place—that is, the reduction in pesticide use—is actually met. By assessing the impact that transgenic pest-management traits had on pesticide use in the United States over a 16-year period (1996–2011), based on data recorded by the USDA, Charles Benbrook (2012) found that while "*Bt* crops have reduced insecticide applications by 56 million kilograms (123 million pounds)," herbicide-resistant GM technology "has led to a 239 million kg (527 million pound) increase in herbicide use in the United States between 1996 and 2011." That is, overall "pesticide use increased by an estimated 183 million kg (404 million pounds), or about 7 percent" (Benbrook 2012, 1). Despite the small, but still beneficial decrease in pesticide use that GM *Bt* crops appear to offer, the advantages seem to diminish when these results are placed in context. If, for instance, GM *Bt* crops are appreciated as pesticide producers themselves—since they have the *Bt* toxin inserted into their genome—then the overall amount of insecticide released in the environment becomes significantly greater as GM *Bt* crops, such as SmartStax GM corn, can produce up to 19 times the amount of chemical insecticide they replace (Benbrook 2012, 7). What is more, other studies demonstrate that the desired decrease in herbicide and insecticide use has been achieved in countries like France, Germany and Switzerland, where GM crops have not been adopted (Heinemann et al. 2014, 78).

Second-generation GMOs: Prospects and problems

The argument for the commercialization of GMOs is built around not only the benefits that this generation of crops already allegedly offers, but also on the fascinating traits that the "second generation" of GM crops will demonstrate. The new generation of GMOs is expected to display a wider range of traits with much broader and more significant applications, such as various forms of disease resistance (e.g., blight-resistant potatoes), various forms of abiotic-stress tolerance (e.g., drought-tolerant GM maize and wheat), other forms of pest resistances (e.g., the Rothamsted aphid repellent GM wheat) and increased nutritional value (e.g., plants that produce healthier vegetable oils with fewer trans fats, "golden rice" to combat vitamin A deficiency and purple tomatoes with beneficial antioxidants) (Agricultural Biotechnology Council 2012a, 6; Baulcombe et al. 2014, 20–27). While GM crops with such traits certainly appear as promising solutions to the challenges of food security, global hunger, malnutrition, climate change and so forth, the reality is that they have been "in the pipeline" for almost two decades and the agbiotech industry has not managed to deliver such crops as promised.

The much-publicized GM "Golden Rice,"[5] for example, which has been promoted in developing countries as a crop that can help fight blindness, is still not available in the marketplace. While the STC has blamed anti-GM sentiment—mostly prompted by Greenpeace campaigns—for the delay of GM Golden Rice commercialization (House of Commons Science and Technology Committee 2015a, 67–69), with a closer look, it becomes evident that this is certainly not the case. On the contrary, it is the results of field trials and laboratory research that have actually blemished the prospects for this "miracle crop." In February 2013, the International Rice Research Institute (IRRI)—the coordinating institution for the Golden Rice Network, which has been working to develop Golden Rice with national partners since 2006—issued a statement "clarifying recent news about Golden Rice." IRRI felt compelled to "clear up two potential misunderstandings" following two articles published in *The Guardian* (McKie 2013) and *Project Syndicate* (Lomborg 2013), which blamed opposition to GMOs and the politicization of the authorization process as the reasons behind the 12-year delay. Setting the record straight, the IRRI announced that, contrary to the claims that Golden Rice soon would be grown in the Philippines, the GM crop "will not be available for planting by farmers in the Philippines or any other country in the next few months, or even this year," as two seasons of field trials in that country had only recently finished. What is more, arguing against the headline of *The Guardian* article, which linked Golden Rice with reduction in blindness in developing countries, the institute asserted that "it has not yet been determined whether daily consumption of Golden Rice does improve the vitamin A status of people who are vitamin A deficient and could, therefore, reduce related conditions, such as night blindness. If Golden Rice is approved by national regulators, Helen Keller International and university partners will conduct a controlled community study to ascertain if eating Golden Rice every day improves vitamin A status" (IRRI 2013). While, in this statement, IRRI claimed that the process "may take another two years or more" (ibid.), in March 2014, the two-year time-frame was apparently discarded as too optimistic. Despite the fact that "beta carotene was produced at consistently high levels in the grain," it was also found from the multi-location field trials that "yields of candidate lines were not consistent across locations and seasons" and, therefore, the com-mercialization of Golden Rice was pushed indefinitely into the future, as it will "only be made available broadly to farmers and consumers if it is suc-cessfully developed into rice varieties suitable for Asia, approved by national regulators, and shown to improve vitamin A status in community conditions" (IRRI 2014). For the time being, the World Health Organization (WHO) opts for more readily available and economical methods to combat Vitamin A Deficiency (VAD) by providing vitamin A supplements, encouraging moth-ers to breastfeed and people to grow fruits and vegetables in their home gar-dens (WHO 2016).

But Golden Rice is not the only second-generation GMO that has not lived up to expectations. As we have seen in this book, the Rothamsted GM Wheat—widely endorsed by GM advocates as "a good example of ground-breaking UK agricultural biotechnology research" (Agricultural Biotechnology Council 2012a, 7)—left the research team "definitely disappointed" by the data gathered from the field trials (Rothamsted Research 2015). Other crops that spearheaded the GM campaign are quite far from having secured the desired results. The blight-resistant GM potato is still in the process of field trials; the GM tomato with high concentrations of health-promoting flavonols and flavonoids is only grown in the greenhouse, while omega-3 enriched GM oilseed rape is at least a decade away from becoming available (Baulcombe et al. 2014, 22–23).

Putting aside the dubious benefits of the current and future generations of GM crops, it is also the broader arguments that do not seem to be well-founded. The lofty aspiration that GMOs may help combat the problems of climate change, global hunger and malnutrition (EuropaBio 2012d, 41, 47) is often construed—outside pro-GM circles—as a "myth" that has nothing to do with research findings (Toke 2004, 1–30). The International Assessment of Agricultural Knowledge, Science and Technology for Development (IAASTD) Global Report (2009)—sponsored by the United Nations and the World Bank, compiled by 400 scientists and approved by 58 countries[6]—did not find GMOs as a credible solution to such global challenges but, as "only one component of a wider strategy including conventional breeding and other forms of agricultural research to provide a series of structural, regulatory, and economic evaluations that relate economic, political, and scientific context of GE crops to their region of adoption" (IASSTD 2009, 95). The reasons for this policy implication were based on the remark that there is "a gap between the reality of how a technology is used (taken up in a given social context) and its 'in the box' design" and the "significant drawbacks" related to GM tech-nology, such as "environmental risks" and "widening social, technological and economic disparities" (ibid.).

Scientific discord: Backward and forward

It takes a very selective reading of the situation to argue that the scientific community, by and large, unanimously agrees on the safety and benefits of GMOs, as there appear to be disagreements on a series of central issues. In fact, scientific divergence is not limited to the areas discussed above, but involves virtually every single aspect of GM technology. A case in point is the *GMO Myths and Truths* report (Fagan, Antoniou and Robinson 2014), which refers to over 600 studies and reports—280 of them being peer-reviewed papers—and casts doubt on the favorable findings for GMOs by underscoring the erratic nature of the technology. The appropriation of scientific evidence to deconstruct the other side's "myths," as opposed to using their own "truths," is a very common practice in the GM controversy—see, for example, the *Science not Fiction* project of EuropaBio (2013c, 20–23), and the corresponding *Twenty Years of Failure* report published by Greenpeace (2015), which is very telling of the current deadlock in the GM debate.

In order to explore the potential of moving away from the apparent impasse that the scientific community has reached, it may help first to examine the reasons why such disagreements exist. Ozawa and Susskind (1985) discern four leading causes behind scientific discords: *miscommunication, differences in the design of inquiries, errors in the inquiry* and *differences in the interpretation of the findings* (27–30).

Quite often, what appears to be "substantive" conflict among scientists is, in fact, a problem of *miscommunication*. This illusion of controversy can be created either intentionally or unintentionally. Miscommunication becomes intentional when scientists use rhetorical devices to sway public opinion toward their own convictions. A common rhetorical scheme employed by both sides to disparage contending positions is the reference to instances of "no evidence" or "inconclusive findings." In the case of GM debate this is particularly common, as scientists who support the technology propound the safety of GMOs by alluding to the fact that there is "no evidence" to the contrary (The Royal Society 1998, paras. 3, 4; 2002, 3, 6, 8, 9) while those who adopt a more skeptical stance stress the fact that there have been no epidemiological studies on the effects of GM food consumption on human health (Hilbeck et al. 2015, para. 2). While such statements may give the impression that there is scientific disagreement on crucial facts, in reality, scientists may agree that there is no conclusive evidence to support either scientific claim. The impression of disagreement may also be given when scientists present the same facts in different ways. An interesting example is how Genetic Modification itself is communicated to the public. GM advocates tend to promote plant biotechnology as a natural extension of traditional breeding methods,[7] which is "just the latest evolution in mankind's never-ending quest to improve how we produce an abundant and safe food supply" (CropLife International 2016). On the other hand, GM skeptics claim the exact opposite by arguing that "technically speaking, the GM transformation process is radically different from natural breeding" (Fagan, Antoniou and Robinson 2014, 25). It is rather obvious that scientists are not disputing the facts on what genetic transformation entails as a process, but are simply using different ways of framing the technology within the wider course of agricultural evolution.

Scientific disagreement may also be a corollary of *differences in the design of inquiries*. While a specific "scientific method" may be accepted as a technique for testing the validity of a given proposition, there are certainly elements of subjectivity involved in the process. These entail, for instance, the framing of the hypothesis, the specification of assumptions and the choice of data (Ozawa and Susskind 1985, 28). It can be the case, therefore, that due to different choices in these three areas, scientists may appear to be in disagreement while they are in reality "talking past each other" (Ozawa and Susskind 1985, 29). A case in point is the disagreement on whether GM *Bt* maize poses a risk to non-target organisms such as the monarch butterfly, a species that has acquired some sort of "totemic significance" in the debate (Perry 2010). While it is asserted in the Baulcombe report that peer-reviewed work on the issue has "concluded that populations of monarch butterflies would not be significantly affected by the cultivation of this GM maize" (Council for Science and Technology 2014, 37), other studies claim that GM pollen certainly harms monarch larvae (Losey, Rayor and Carter 1999). According to Professor Perry, "the Baulcombe example of monarch butterflies is naïve" as it ignores the "clear evidence of some risk to non-target Lepidoptera from GM *Bt* maize" (Perry 2014, para. 7). The disagreement on this crucial issue seems, however, to be mitigated when one considers that the study to which Baulcombe refers (Sears et al. 2001) concerns "a North American species from outside the EU receiving environment," while the studies of (Perry, Devos et al. (2010), which show impact on the mortality rate of Lepidoptera have used species native to Europe.

Erroneous scientific findings are the third cause of conflicting evidence that can spark policy debates (Ozawa and Susskind 1985, 29). In this case there is scientific disagreement on the validity of a given body of evidence and, in order to overcome this controversy, the scientific community usually demands that the study be replicated.

The fourth and final major reason for scientific disagreement is the *differences in interpretation of findings*. Uncertainty represents perhaps the major interpretive challenge for scientists, in particular for those working at the frontiers of a technoscientific field (Ozawa and Susskind 1985, 29, 30) such as that of

GM technology. In those instances in which the efficiency of a new technique is tested or the toxicity of a novel GM crop is examined, results cannot be interpreted according to a predetermined and consensual scientific protocol but are contingent on the scientists' own subjective reading of the results.[8] A highly publicized example of the divergent interpretation of findings has been the work of Séralini (de Vendômois et al. 2010). After obtaining raw data of 90-day feeding regulatory trials on rats eating GM corn or soy data—turned over by court order, as the data were formerly kept secret—Séralini and his colleagues conducted their own assessments in order to compare the results with the data previously obtained. Significant statistical differences were found (approximately 9 percent) regarding mostly kidney and liver functions. The official approval committees and the companies producing these crops, however, interpreted these results as "irrelevant [to] the safety of GMOs" (de Vendômois et al. 2010, 591).

The suggested typology, it seems, is significant not only for the analytical rigor it offers but, most importantly, for the implications it has for practical policy. It could be argued, for instance, that in a mediated environment of dialogue, where disagreeing parties are invited to participate voluntarily, the magnitude of differences may be significantly reduced so long as these four causes of discord are addressed. Miscommunication and rhetorical devices may be directly addressed through controlled dialogue, while tentative agreement could be reached on the design of inquiries and how results are interpreted in existing or future assessments. This policy initiative is further discussed in the final section of this chapter. Before doing so, however, we have to assess whether scientists indeed offer objective and value-free opinions or whether they are equally prone to producing results that are in accord with the interests of affiliated stakeholders and their own subjective worldviews.

Scientists, vested interests and ideology

In Chapter 4 we argued that scientists do not theorize, perform experiments and invent in a social and natural vacuum. Rather, like all social actors, they act within a particular environment of action that includes networks of position–practice relations, rules and resources. The actions of scientists, it was suggested, are contingent not only on the structural environment but also on their own conjuncturally specific knowledge of the situation, deeply seated dispositions and active agency. The subjectivity underlying scientists' actions was theoretically introduced through the TAI-SST scheme and acquired a more substantive dimension in the following chapters where specific examples were given of the ways that personal convictions and motives affect scientists' actions, which may sometimes deviate from normative expectations. During the discussion of the precautionary principle, it was also argued that, as a process, risk assessment is not immune to social influences, since normative arrangements influence the way the evaluation is framed and carried out (Wynne 1992). We can now turn to the illustration of specific examples from the field of GMOs, which deconstruct the conception that scientific work is insulated from economic and political interests, emotional outbursts and personal idiosyncrasies.

To reiterate, the choice of assumptions, methods, models and measuring devices that a researcher makes is not rigidly defined by objective practice (Ozawa 1996, 226) as it is influenced by social, economic and political factors such as the researcher's institutional affiliation, source of research funds and disciplinary training (Knorr-Cetina 1982). Institutional affiliation and source of funding—or, called differently, the problems of conflict of interest (COI) and revolving doors—do not seem to influence only the impartial functioning of organizations like EFSA, as we saw in Chapter 6, but also the results of peer-reviewed articles. To be more precise, in a systematic review of the scientific literature on the health risks or nutritional assessment studies of GM products, Diels (2011) noticed "a strong association [...] between author affiliation to industry (professional conflict of interest) and study outcome" (197). More precisely, it was found that from all 44 papers with identified financial and/or professional conflict of interest (COI), 43 contained a favorable outcome for GMOs, and only one unfavorable. Conversely, from 37 papers with no identified COI, 27 reached a favorable outcome, eight unfavorable and two a neutral (Diels et al. 2011, 200). That is, there is only a 2 percent chance for scholars with COI to reach an unfavorable outcome for GMOs while without a COI the probability rises to 22 percent.

Emotional outbursts and dogmatic behavior do not only characterize anti-GM zealots, as some would call them, but also prominent figures advocating GMOs. The decision of the Scottish government to ban cultivation of GM crops in their territory, for instance, caused a considerable stir among some divisions of the scientific community. A few days after the formal announcement, the former CSA to the Scottish government, Professor Muffy Calder declared she was "disappointed and angry" as the ban could have "apocalyptic" effects. More precisely, Calder argued that as a result of the GM ban, Scottish crops could be exposed to diseases that "could come and wipe us out in an apocalyptic sense" (qtd. in Amos 2015). Further reactions were voiced by a cluster of research centers and professional associations—including among others Rothamsted Research, the European Academies Science Advisory Council (EASAC), the John Innes Centre, the Royal Society, the Science Council, Roslin Institute (creators of Dolly the Sheep) and the National Farmers' Union—spearheaded by the charity Sense about Science, which sent an open letter to Rural Affairs, Food and Environment Secretary Richard Lochhead in an effort to warn of the "negative effect on science in Scotland" this decision would have (Sense about Science 2015). In the letter to Lochhead, Sense about Science characterized the decision as "political and not based on any informed scientific assessment of risk" and in dramatic tones asked "urgently for a meeting where researchers can discuss these concerns with [Lochhead] and consider ways to protect the freedom and integrity of science, and its use in policies, in Scotland in the future" (ibid.). The appeal was endorsed by Mark Lynas, the political director of the Cornell Alliance for Science (CAS)—a program launched in 2014 with a $5.6 million grant from the Bill & Melinda Gates Foundation and a goal to "depolarize the charged debate" about GMOs (Malkan 2016). In a *New York Times* op-ed, Lynas argued that Scotland, together with the other 16 EU countries who opted out of GM cultivations, formed the "Coalition of the Ignorant" by following a policy analogous to Europe's prohibiting the printing press in the 15th century and which aligned the EU with some "unsavory allies" such as Russia and Zimbabwe (Lynas 2015).

In all three instances, the issue of GMOs was framed once again within the typical binary "science versus politics" scheme, presented as a case of political interventionism banning technoscientific artifacts that are unanimously deemed safe and beneficial. Responding to these instances of animosity, Professor Brian Wynne described the pro-GM stance held by some of his fellow academics as a "religious crusade" touting visions of "damnation."

> The GM debate is not black and white. It's not just a binary option— either we have GM crops or we don't. It's far more complicated than that. The kind of language that has been used in the past few days, not just by Muffy Calder but also in the letter from Sense About Science, it's like an obsession with GM. It's a bit more like a religious crusade. The idea that if we don't have GM then somehow it's dereliction and hell and damnation and starvation is total rubbish. (qtd. in Amos 2015)

The fact that scientists are "just emotional as anyone else" (Wynne, qtd. in Amos 2015) can account for not only the above instances of personal acrimony but, most significantly, for the stalemate in meaningful scientific dialogue. We have already seen, for example, how Take the Flour Back and Rothamsted Research did not manage to bring independent experts to the table and have a "public debate" on GMOs amidst a tense atmosphere of mutual threats and vilifications. We have also discussed the unprecedented reactions of certain distinguished members of The Royal Society toward the study of Ewen and Pusztai who, instead of engaging in systematic dialogue on the design and findings of the study, preferred to break the norms of anonymity and impartiality in an effort to have the study retracted; even though it had successfully passed the peer-review process of *The Lancet*.

A recent failed attempt to promote public scientific debate is very indicative of the barriers of distrust that seem to divide segments of the scientific community. In October 2014, the EU-funded research project GMO Risk Assessment and Communication of Evidence (GRACE) published the results of its 90-day feeding studies with GM MON810 maize, which aimed at testing the added scientific value of feeding trials for risk assessments of GM plants. This was a project with massive scientific significance, as the importance of the feeding studies has been stressed by Séralini (Séralini et al. 2012) and other cluster 3 scientists, while at the same time have been deemed unnecessary by Perry and the EFSA GMO Panel, as already discussed. The same month the results were published, GRACE invited experts and interested members of the public to participate in a scientific discussion of the research findings, which showed that "food containing up to 33 percent MON810 maize did not have any negative effects on male or female Wistar Han RCC rats following sub-chronic exposure (90-day feeding trial)" (GRACE 2015b). For the purposes of public participation, an open discussion forum was set up, hosted by the scientific journal *Archives of Toxicology*. Since the study was of high importance for how GM risk assessment is carried out in the EU, GRACE was committed to conducting research "in a systematic, transparent and inclusive way" (GRACE 2012).

Nonetheless, from the moment the research findings were published in the peer-reviewed journal, Testbiotech (an association of experts promoting industry-independent research and public debate on the impact of biotechnology) and its executive director, Dr. Christoph Then—whose involvement in the GM debate we discussed in Chapter 7—fiercely criticized the study. In the exchange of open letters between the two parties, Testbiotech—by alluding to the external expertise of a toxicologist who remained anonymous throughout the correspondence—suggested that the study endorsed by GRACE should be retracted, as its findings were scientifically flawed and the "outcome was manipulated" because the project's overall transparency was questionable for unclear COI between the journal and industry (Bauer-Panskus and Then 2014). Although Professor Pablo Steinberg, one of the main authors of the GRACE paper, replied to each scientific point raised by Testbiotech (Schiemann 2014), and the EC also confirmed, for its part, that GRACE "engaged in a wide stakeholders consultation for the design and execution of the study but also analysis and interpretation of the results" (European Commission: Directorate-General for Research and Innovation 2014), Testbiotech and Then remained unconvinced. GRACE invited more than 700 stakeholders and scientists to take part in workshops and submit written comments, including Testbiotech, but apart from the mentioned correspondence, neither Then nor any other representative of the association engaged in dialogue. Instead, they found it more sensible to use media outlets to discredit project GRACE and use as evidence the analyses carried out by the anonymous toxicologist (GRACE 2015b).

Reframing the GM debate: Ecological Disputes and Democratic Trajectories

It has been argued thus far that GMOs are neither unproblematic technoscientific products nor victims of political interventionism. Rather, they are part of a rich and diverse conjuncture that has divided scientists on their alleged risks and benefits, and political actors on the efficiency of regulatory systems and policies that need to be shaped. If this argumentation is accepted, then two fundamental questions need to be answered: What is the GM controversy? How can we move beyond the currently fragmented state of affairs?

The GM controversy as an embracing ecological dispute

While among European citizens "there is widespread support for medical (red) and industrial (white) biotechnologies," there is "general opposition to agricultural (green) biotechnologies in all but a few countries" (European Commission 2006, 3). Looking across public perceptions of different types of technologies, resistance to GMOs is the exception rather than the rule, as there "is no evidence that opposition to GM food is a manifestation of a wider disenchantment with science and technology in general" (ibid.). In general, Europeans are not in favor of development of GM foods, as they do not see any benefits and consider them probably unsafe or even harmful (European Commission 2010a, 7).

One major issue that distinguishes green biotechnology from other biotechnological innovations is its core environmental component—a component that often appears to play a peripheral role in how the debate is framed. Once released into the environment, GMOs become parts of the local ecosystems and food webs and can affect both the natural environment and the social order in manifold ways (Kinchy 2012). For this reason, as we have already discussed, the GM controversy functions as a lightning rod for broader economic, political and cultural concerns. Therefore, if we want to break away from a dualistic science/politics scheme and reconsider the dynamics of the GM controversy, we have to take into account both the central role that the environmental component plays and the substantial involvement of science, politics, economics and culture.

Gunnar Sjöstedt's (2009) contribution from the nascent and multi-disciplinary field of conflict resolution can prove of great relevance to the task at hand. For Sjöstedt, ecological disputes or conflicts may arise as a result of environmental problems triggered or influenced by human activity. Such activities may include agricultural practices, generation and disposal of hazardous waste, transportation of goods and people, or war. The dispute or conflict that may erupt due to anthropogenic environmental problems may be of low, moderate or high intensity. Depending on the degree to which the environmental component of an ecological disagreement is linked to other contentious issues, three broad types of conflict can be distinguished: *pure ecological conflicts, embedded ecological conflicts* and *embracing ecological conflicts*. Pure ecological conflicts are dominated by one specific environmental issue—although other matters of secondary importance may be involved—and usually represent risk rather than a crisis. An example of such a conflict is the dispute between Austria and the Czech Republic for the Temelin nuclear plant, which Austria wants closed on account of a nuclear accident, and the Czech Republic needs to be operational for the production of electric energy. Embedded ecological conflicts refer to disputes in which the impact of the environmental issue is hard to determine, as it is intertwined with other matters of a different nature. Pollution in a river such as the Jordan or the Nile, or an international lake such as the Caspian Sea, can be seen as an illustration of an embedded ecological conflict, since the issue of water pollution is often embedded by societies in a broader agenda that includes natural resources, human health and transportation of goods and people. In embracing ecological conflicts, too, a multitude of issues is involved, but in this case, the environmental component is dominant and embraces the other issues. Illustrations of embracing ecological conflicts include the negotiations on climate change and ozone layer depletion. In these two cases, the talks have a clear focus on the environmental issue, but other concerns are also part of the problematic, such as land use, industrial production, energy and more (Sjöstedt 2009, 227–29).

In light of the above threefold typology, we can argue that the GM controversy may be classified as a *low-intensity embracing ecological dispute*.[9] Through this framing, the environmental aspect of GM technology is acknowledged as the dominant component of the dispute, but the other involved issues also are given due attention. The reframing of the debate in these terms can help bring to light the rich and diverse nature of the conjuncture by exploring its overarching distinctive characteristics. For Sjöstedt, environmental issues share seven general features that tend to shape how an ecological conflict is addressed and handled. While the author primarily refers to instances of pollution, I think the case of GMOs dovetails with the interesting analysis he offers. I discuss each point by making connections to the GM controversy.

- *They have a special trans-boundary character.* "The trans-boundary dimension of an environmental issue can be conceived of as a package of relations of interdependence linking two or more countries" (Sjöstedt 2009, 231). In the case of GMOs, the trans-boundary character they acquire is of crucial significance. This refers not only to the fundamental issue of gene flow— where GM crops can cross-pollinate with their wild relatives resulting in hybrid crops—but also to the fact that once released into the environment GM material can find its way into numerous parts of the food web. For the minimization of cross-pollination, appropriate technical and organizational measures need to be taken during cultivation, harvest, transport and storage (European Commission 2009). Since, however, GMO admixtures are in essence unavoidable, "GM-free" food can no longer exist in an absolute sense (GMO Compass 2006). In order to safeguard coexistence between GM and non-GM crops, protect consumer sovereignty and mitigate the magnitude of adventitious gene flow, the EU has devised a very thorough traceability and labeling framework. Traceability enables tracking GMOs and GM food/feed products at all stages of the supply chain and, when the proportion of GM ingredients is higher than 0.9 percent, labeling requires clear indication of food/feed products (European Commission 2015d). Unintentional transboundary movement of GMOs is highly undesirable in the EU as under EC Regulation 1946/2003 member states are advised to "take appropriate measures to prevent unintentional transboundary movement of GMOs." In the event that GMO movement is noticed, the member state shall "without delay consult the affected or potentially affected States to enable them to determine appropriate responses and initiate necessary action, including emergency measures in order to minimize any significant adverse effects" (European Commission 2003c, L287/6).
- *They are issues of high complexity.* Another characteristic that environmental issues share is the difficulty they pose for efficient analysis, as they usually require multilayered knowledge on a variety of closely related issues (Sjöstedt 2009, 232–33). The problem with issue complexity is that it "may lead to an asymmetrical distribution of knowledge/information with the effect that awareness and understanding of a disputed issue varies considerably across stakeholders impeding joint perception, the construction of consensual knowledge and effective problem solving in a process of conflict resolution" (Sjöstedt 2009, 234). The efficient analysis of GMOs indeed presents significant challenges to the scientific community, as risk, hazard and uncertainty are part and parcel of the plant's Environmental Risk Assessment (ERA)

(Perry 2014, para. 3). EFSA identifies seven specific areas of concern that ERA has to address thoroughly: persistence and invasiveness of the GM plant; plant to micro-organism gene transfer; interaction of the plant with target organisms; interaction of the plant with non-target organisms; impact of cultivation, management and harvesting techniques; effects on biogeochemical processes; and effects on human and animal health (EFSA 2015a). As already discussed, the scientific community is divided on virtually every single one of the areas identified, and consensual knowledge is still not a reality in the case of GMOs.

- *They tend to involve a complex combination of participants in conflict resolution.* Instead of being confined to disagreements between states and governments, disputes instigated by environmental issues—due to their high technical complexity—have the distinct characteristic of requiring the presence of scientific expertise and attracting the attention of NGOs and green groups (Sjöstedt 2009, 234–35). The involvement of politicians, scientists, NGOs and environmental groups in the GM controversy has been demonstrated throughout this book as giving the debate its own colors and contours.
- *Their distribution of negative rather than positive values.* In ecological conflicts, positive values are associated with the natural resources that will be preserved (e.g., fresh water) by the abatement of an environmental problem (e.g., pollution), while negative values are represented by abatement costs. In these types of situations, environmental deterioration is construed as a cost that has to be mitigated by another cost generated by the measures taken in order to achieve abatement. As a result, involved parties typically negotiate or argue about the distribution of such negative values and the trade-off between environmental destruction costs and environmental abatement costs (Sjöstedt 1993; 2009, 235). While the GM discussion does not revolve around environmental destruction and the distribution of abatement costs, there is a strong tendency for both sides to stress negative rather than positive trajectories associated with the presence/absence of GMOs in the environment. GM advocates argue, for instance, that without GM crops it will be tough to tackle the global challenges of food security, world hunger, climate change, population growth, land use change and so forth. While not all GM advocates would share the former Scottish CSA's "apocalyptic" tone of what lies ahead if GMOs are not widely commercialized, it is very common for undesirable scenarios to be articulated in the introductory parts of scientific reports as opening leads for the need of GMOs (Baulcombe et al. 2014; BBSRC 2009; Council for Science and Technology 2014; The Royal Society 1998, 2009). The distribution of negative values associated with the presence of GMOs is stressed even more passionately by GM skeptics who argue such technological products create irreversible damage to the

environment, may harm human health and in essence pave the way for a few agbiotech corporations to control the food chain and phase out small farmers (Fagan, Antoniou and Robinson 2014; Friends of the Earth International 2012; Greenpeace 2015). From this point of view, negative values are distributed even among parties who choose not to adopt GM crops or foods but, nonetheless, are recipients of such results due to the choices of others.

- *They are plagued by multi-layered uncertainty problems* (Sjöstedt 2009, 235). The claims of the *Baulcombe Report* that there are no "unknown unknowns" in GM plant breeding (Baulcombe et al. 2014, 38) can now be characterized as overly optimistic. New scientific knowledge and innovative technologies such as GMOs entail different aspects of incertitude, such as risk, uncertainty, ambiguity, ignorance and indeterminacy (Felt and Wynne 2007, 31–42). The fact that virtually every aspect of GM technology is contestable— ranging from the mutations that DNA insertion may cause. to the reduction in the use of pesticides and from the increase in yields to the harm on non-target monarch butterflies—attests to the fact that multilayered uncertainty problems are intrinsic to the GM debate.

- *Their propensity to be framed as either crisis or risk.* Environmental issues in a potential or actual ecological dispute tend to be framed either as crisis or risk, depending on whether real and considerable damages have occurred (Sjöstedt 2009, 236). The choice of appropriate framing is contingent, therefore, on the perspective of the stakeholders involved. From a purely technocratic perspective in the EU, GMOs may be framed as posing a potential risk, as the terms risk assessment and risk management self-evidently suggest. From the point of view of non-GM alfalfa farmers in Washington State, however, who are now unable to export their crops as the result of transgene flow, the uncontrollable presence of GMOs may be framed as a crisis that has caused significant environmental and economic damage (Graham 2016).

- *They tend to be securitized.* Securitization of environmental issues occasionally appears by some, or even all, of the stakeholders in an ecological dispute. When this happens, the dispute tends to intensify as certain parties become more willing to take risks and resort to unilateral strategies (Sjöstedt 2009, 237–38). This pattern has been observed, not only in the more radical groups that "decontaminate" field sites on the grounds that GMOs pose serious threats to human health and the environment, but also across authorities in the EU. A case in point is the "GMO-free Europe" initiative— supported by farmers, agronomists, grain traders, regional administrators and politicians from all over Europe—which vows to establish "GMO-free regions" in Europe so as to "protect conventional and organic seeds" from GM contamination (GMO-free Europe 2005) and ensure that "biodiversity, independent farming, and regional quality food production" are not eroded by "the exclusive control of seed by fewer and fewer companies" (Levidow and Carr 2010, 240).

Public involvement: Backward and forward

If the suggestion that the GM controversy should be conceptualized as an embracing ecological dispute is sustained on a theoretical level, then implications for practical policy emerge. Since, for all three types of ecological conflict, the optimal resolution lies in a mediated environment of dialogue and deliberation (Sjöstedt 1993; 2009, 239; Susskind, Moomaw and Gallagher 2002), there is no particular reason why such kinds of mediation are not also applied to the GM dispute. To this end, in this final substantive part of the book, I will flesh out an alternate model of inclusive dialogue that can complement formal decision-making frameworks. This model is a critical synthesis of different types of dialogue, practiced in the resolution of long-standing conflicts and environmental disputes, and is suggested as a tentative trajectory toward more inclusive technoscientific progress and democratic policymaking.

Before doing so, however, it must be clear from the outset that it lies well beyond the scope of this book to offer a comprehensive analysis of the intricate and, at the same time, subtle issue of public engagement and the manifold forms this may acquire. Nonetheless, a fundamental discussion needs to be made, and for two reasons. First, the desirability and conditionality of public engagement were broached, not only during the discussion of the precautionary principle, but have been a recurring theme throughout this book which, nonetheless, has not been explicitly deliberated. As a result, for the sake of argumentative clarity, specific answers need to be given as to what public involvement entails, how it is implemented, under what conditions it becomes substantive and in which ways it can resonate with scientific progress and the social order. Second, the suggested reframing of the controversy has raised a variety of issues regarding decision-making practices and policy formation processes. I start with the exploration of the first theme and then move on to offer some tentative suggestions as to how this book's holistic framework can potentially help overcome substantive differences among stakeholders.

Is democracy unscientific?

The second constitutive element of the precautionary principle, mentioned in the previous chapter, is the presence of a cost–benefit analysis in the decision-making process. As we have already discussed, damages—whether restricted or irreversible—can occur at the level of the cell, individual organism, population or ecosystem. Consequently, the effects of such damages may be biological, ecological, social, economic or cultural; they may be distributed equally or disproportionately among individuals, populations or geographical areas, now or in the future. "Because systems are complex and outcomes are not always predictable, it becomes extremely important for decision makers to specifically identify the parameters that are used to assess the potential effects of a proposed activity" (Schettler and Raffensperger 2004, 67). Therefore, since the commercialization of GMOs has *de facto* ramifications (whether benign or malignant) on populations and ecosystems, it is suggested that policymakers take into account all parameters and respect the needs of all parties involved. In fact, the EU has clearly pledged itself to open and democratic decision-making processes. "The use of biotechnology is [however] not without controversy and the enhanced use of biotechnology needs to be accompanied by a broad societal debate about the potential risks and benefits of biotechnology including its ethical dimension" (European Commission 2007, 2). In the report, *Life Sciences and Biotechnology—A Strategy for Europe*, it is openly stated:

> Societal dialogue and scrutiny should accompany and guide the development of life sciences and biotechnology. Life sciences and biotechnology should be developed in a responsible way in harmony with ethical values and societal goals […] Dialogue in our democratic societies should be inclusive, comprehensive, well informed and structured. Constructive dialogue requires mutual respect between participants, innovative approaches, and time. (European Commission 2002, 11–12)

While this is so, more often than not there are either mild implications or unequivocal statements that such democratic practices do more harm than good. As with the issue of risk avoidance, with some social actors the precautionary principle proves to be an unfavorable approach on the grounds that the involvement of non-specialists within the decision-making process seriously jeopardizes the prospects of financial growth and technological development. In this way, the argument goes, the "anti-technology nature of the precautionary principle" is legitimized and "research into technologies that can improve our safety and well-being" is eventually dropped (Conko and Miller 2001, 303).

One reason why public involvement with technological developments has negative connotations is grounded in the concern among policy, scientific and industrial elites about the alleged widespread public uneasiness with science. "Indeed, fear is expressed that this may be just the beginnings of a more paralyzing reaction against technoscientific innovations essential for Europe's

survival in the face of accelerating global competition. The place of science
[…] as a key agent of governance and government, able both to enlighten, and
to generate public legitimacy for democratic policy commitments, is seen as
seriously weakened by this public unease" (Felt and Wynne 2007, 13). This is
also clearly seen in the *Life Sciences Strategy* report: "Whilst confirming and even
emphasizing the importance of dialogue, however, the Life Sciences Strategy
seems to demonstrate a frustration that the economic potential of the life sciences is not being achieved because of public mistrust" (Lee 2008, 80). The
same ambivalent stance on public opinion was also recently expressed by MP
Owen Paterson in his speech at the UK–Ireland Food Business Innovation
Summit:

> I sympathize with the incredibly difficult position the Commission are in
> given wildly conflicting views on GM across member states. The EU has
> the strongest and strictest safety-based regime for GMOs in the world—
> and it's right that products should be subject to such controls. *But there is
> more the EU as a whole can do to facilitate fair market access for products which have
> been through that system. The EU is being left behind when it comes to GM, and
> I fear we'll regret it if we don't try and catch up.* (Paterson 2013, stress added)

In more extreme cases, such as the controversy surrounding the mandatory
character of GM labeling, the potential contribution of consumers and the
public is de facto belittled by certain proponents of GMOs. Social actors
(including agribiotech companies, scientists and some NGOs) who oppose
obligatory GM labeling do so on the grounds "that food labeling would be
confusing, misleading and irrelevant for consumers […] as *GM labels inform no one* and
they diminish consumer choice" (Klintman 2002, 74–75, stress added). If
the public is construed as *in principle* incapable of comprehending GM labeling, there is no doubt that for these social actors inviting nonspecialists to the
decision-making process would be nothing less than unthinkable. This has also
been the unequivocal stance of Dick Taverne—then chair of the Sense about
Science charity which, as we saw earlier in the chapter, openly supported the
Rothamsted Wheat trials—and has argued against "the fashionable demand
by a group of sociologists for more democratic science, including more
'upstream' engagement of the public and its involvement in setting research
priorities" on the grounds that "science, like art, is not a democratic activity.
You do not decide by referendum whether the Earth goes round the Sun"
(Taverne 2004, 271). Whether out of temporary fear, ignorance or deeply
seated dispositions, the public is often considered to be very prone to anti-
scientific sentiments and should, the argument goes, be—either de facto or
"in the last instance"—excluded from policymaking. Policymakers can readily

fall into the trap of seeing civil society as "outsiders, to be taken into account, for sure, but as 'irrational,' prone to be scared without reason, and always to be monitored by opinion polls" (Felt and Wynne 2007, 25–26). I spend the remainder of the section trying to demonstrate the logical fallacies of this argument and offer a more positive note as to how public involvement can be very beneficial to technological innovation.

The first problem with the argument against public participation is that it appears to be a corollary of technological determinism. While this idea has been extensively discussed in Chapter 2, but also schematically referred to throughout the book, I would like to discuss very briefly how technological determinism does not allow any bridge building between technological innovation and public involvement. The insistence of certain social actors—such as agribiotech corporations, politicians scientists and so on—that the public should be pro-science and pro-innovation removes, in effect, any discussion about what kind of science and innovation the public needs or wants. The public is asked unconditionally to embrace science, technology and innovation without questioning where these lead. This message was perfectly crystallized in the "Science Matters" speech Tony Blair gave to the Royal Society in 2001:

> The idea of making this speech has been in my mind for some time. The final prompt for it came, curiously enough, when I was in Bangalore in January. I met a group of academics, who were also in business in the biotech field. They said to me bluntly: "Europe has gone soft on science; we are going to leapfrog you and you will miss out." They regarded the debate on GM here and elsewhere in Europe as utterly astonishing. They saw us as completely overrun by protestors and pressure groups who used emotion to drive out reason. And they didn't think we had the political will to stand up for proper science. (Blair 2002, 155)

The principles of technological determinism are once again touched upon: (a) There is the belief that science and technology move forward, and society has to adapt accordingly; (b) technological advances ought to be readily espoused unless societies desire to be left behind the evolutionary path. In light of this rhetoric, public involvement is construed as a disturbing factor; people who oppose or question the fruits of technoscience are simply irrational social actors (individual or collective). What happens eventually is: "We get locked into a circular discourse in which the only answer to the question 'What is scientific and technological progress?' is 'whatever our innovation systems are delivering'" (Wilsdon, Wynne and Stilgoe 2005, 25–26). However, for many scientists, this is certainly not the case. "Creating something just because we are now able to do so is an inadequate reason for embracing a new technology.

If we have advanced tools for creating novel agricultural products, we should use the advanced knowledge from ecology and population genetics as well as social sciences and humanities to make mindful choices about to how to create the products that are best for humans and our environment" (Ellstrand 2001, 1545). Instead of asking questions of the type: "Why this technology? Is another feasible? Who needs it? Who benefits from it? What will it mean for and my family?" (Wilsdon and Willis 2004, 28), "all too easily, we fall back into a set of polarized debates in which participants are cast as either 'pro-innovation' or 'anti-science' " (Wilsdon, Wynne and Stilgoe 2005, 26). It is not too difficult to see how this theme was present in the GM wheat field trials. On the one hand, the proponents of the field trials repeatedly confirmed their pledge to science, evolution, progress, problem-solving and so forth, while the skeptics were usually portrayed as a group of anti-scientists who were willing to destroy private property in order to stop progress. Therefore, for those who believe that science and technology follow a linear evolutionary path, public opinion is at best unimportant and at worst anachronistic. However, doing away with the "growing uneasiness which affects the relations between science and society" cannot be achieved by excluding civil society from policymaking and mocking public opinion as irrational or anachronistic. Rather, social skepticism "can only be realized through more participatory modes of governance" (Felt and Wynne 2007, 53).

The second logical fallacy of wanting to exclude sectors of society from policymaking is that it espouses some form of *reductionism*; in fact, one can discern a twofold reductionism. One such reductionist mode of thought takes place when cultural, political and ethical parameters are placed in the background of the discussion, and the assessment of technoscientific promises or achievements is replaced with the simple calculus of economic growth (Wilsdon, Wynne and Stilgoe 2005, 26–27). In fact, the motif of measuring the (potential) contribution of science and technology in primarily economic terms has been echoed throughout this chapter in speeches of politicians who clearly opt for GMOs, in reports of corporations that indulge in depicting graphs of extraordinary future financial benefits, and in governmental or supra-governmental documents that allude to the fear of "being left behind." To be sure, assessing economic benefits is an integral part of the discussion on technological advances, and such analyses do not constitute a problem. Issues arise when economic terms are presented as incontestable benefits that rule out the need for public dialogue. The concentration on economics in decision-making can be dangerous on conceptual, practical and ethical grounds, as it tends to hide significant uncertainties and value judgments shared by the public sphere behind a display of apparently inevitable numbers (Lee 2008, 49). This form of thinking usually has a companion reductionist stance—the one that frames all public opinions and concerns in terms of risk (Wilsdon, Wynne and Stilgoe 2005, 27). By conflating public opinion with the notion of risk and mistrust, one does great injustice to the perceptive abilities of citizens. A report compiled by the Expert Group on Science and Governance on behalf of the European Commission stated:

> The public is thought to fear science because scientific innovations entail risk. Both science and risk, however, are ambiguous objects. It is frequently assumed in policy circles that the meanings of both for citizens must be the same as for experts, but that assumption is, in our view, itself a key element in generating "public unease." The widespread sense of unease—sometimes expressed as "mistrust of" or "alienation from" science—must be seen in broader perspective. We conclude indeed that there is no general, indiscriminate public disaffection with nor fear of "science." Instead, here is selective disaffection in particular fields of science, amidst wider areas of acceptance–even enthusiasm. (Felt and Wynne 2007, 9)

To summarize, scientific and technological progress need not be construed as a closed enterprise in which only few exclusive members have the right to voice their concerns or aspirations. Public involvement is not about hampering technological progress, but about expanding the horizons of development by incorporating social and ethical aspects offered by citizens. This inevitably means that non-specialists should be given the opportunity to discuss with scientists, businesspeople, politicians, policymakers and so on to collectively shape the trajectories of technological advances. If the viewpoint that either partially diminishes or totally rejects public involvement in the development and/or commercialization of science and technology is abandoned on the grounds that, at the very least, it is deterministic and reductionist, then the possibility for active civil contribution opens up. But what does this entail?

Public involvement in a nutshell

There is no one ideal process for public participation, but a variety of methods and techniques can be used according to the desired goal. These include citizens' juries and panels, deliberative opinion polls, standing consultative panels, consensus conferences, internet dialogues and focus groups (Kass 2001, 6). During such participatory practices, the essence of public involvement revolves around a transparent, interactive process by which social actors and innovators become mutually responsive regarding the ethical acceptability, sustainability and social desirability of the innovation process and its marketable products (Gee 2013, 661). Transparency demands, among other things, the sincere communication about what is known, what is not known and about uncertainties that exist. This kind of deliberation should involve all interested parties including scientists, regulatory authorities, politicians and the public where alternative trajectories are considered and limits are not placed in advance on the range of alternatives (Hansen and Tickner 2013, 34–35). Having said this, it should be made clear that public involvement is *not* about romanticizing the capabilities or motivations of the civilians engaged and construing the public as social actors full of enthusiasm eager to perform as ideal "participatory citizen-scientists" (Felt and Wynne 2007, 41). The value of involving lay people does *not* lie in the assumption that they are more knowledgeable or environmentally committed, but in the complementary value of their perspectives, which are often broader in scope, more firmly grounded in real-world conditions or more independent than the narrow professional perspectives that can accompany specialist expertise (Gee and Stirling 2004, 107).

In order for civil involvement to be meaningful and substantive, it must not simply inform decisions, it should shape them (Wilsdon and Willis 2004, 39). This presupposes, however, that the objectives of any public engagement should be clear from the very beginning (Wilsdon and Willis 2004, 38) and that it is initiated early in decision-making, beginning with setting goals, when the health and well-being of the public and environment are at stake (Schettler and Raffensperger 2004, 79). If this condition is not met, it is too difficult a task to achieve public involvement once technologies are in the process of commercialization. This is so because in the final stages of technological innovation corporations have already invested huge amounts of capital and, consequently, are very likely to exert mounting pressures so as to meet their scheduled timeframe (Zahabi-Bekdash and Lavery 2010, 248).

There are numerous cases in which public involvement has contributed actively to the acquisition of useful knowledge and helped scientists find solutions to pressing problems. Some examples include AIDS patients (Felt and Wynne 2007, 59), Alzheimer's disease (Wilsdon, Wynne and Stilgoe 2005, 30–32), the relation between asbestos and lung cancer, and the developmental and reproductive harm from exposure to PCBs (Gee and Stirling 2004, 107). Civil participation, therefore, is highly desirable, as it is about inviting scientists to reflect on the social and ethical dimensions of their own work. In this way, scientists acquire a broader perspective regarding the significance of their craft and the ways this can be used or appropriated. Public engagement is clearly *not* about delaying technological developments and is certainly *not* about members of the public standing over the shoulder of scientists in the laboratory, taking votes or holding referendums on what they should or should not be doing (Wilsdon, Wynne and Stilgoe 2005, 35).

The GM controversy: A tripartite involvement

Going back to the GM debate, it has been argued so far that there are three overarching characteristics:

- There is *scientific disagreement* on central aspects of GM technology; the disagreement relates not only to the environmental component of the technology but is *multi-disciplinary*, as it entails economic, political, social and cultural issues;
- There is *significant pressure on the political field* for more robust regulation of GMOs from both GM advocates and skeptics; with the former demanding a more flexible and expedient regulatory procedure and the latter arguing for an inclusive framework which, at the very least, is informed by the precautionary principle;
- There is *public distrust* specifically toward agbiotech, while there is widespread support for other applications of biotechnology in the EU.

Ozawa and Susskind suggest that disputes that entail scientific evidence typically involve three sets of actors: *affected interests* (individuals and groups likely to receive the benefits or to bear the costs of a particular policy decision; these may include consumers, farmers, residents in areas of GM cultivation, corporations investing in the technology, environmental NGOs, professional associations and so forth); *decision-makers* (elected or appointed officials with decision-making authority); and *scientists* (technical experts called upon to provide pertinent technical expertise) (1985, 26). It is of particular importance to note that these actors are rarely mutually exclusive or unaffected from one another's interests and motives. Public officials, for instance, can hold allegiances to agencies with a stake in decisions and become "affected interests," such as the instances of COI and revolving doors we have discussed. Scientific experts may also be linked with an affected interest through the source of funding, or individuals who are directly affected by a decision may also be professionals with expertise.

We are, therefore, witnessing a tripartite situation in which each side is in discord with the others. What is more, the argument that the scientific community can arbitrate this dispute cannot be sustained as: (a) there is disagreement among scientists as well, and (b) scientists themselves can be "affected interests."

Formal frameworks of inclusive governance

Two recently developed and very promising models aimed at precisely bridging the differences between decision-makers, affected interests and scientists deserve closer scrutiny: the International Risk Governance Council (IRGC) Risk Governance Framework (IRGC 2016b) and the *General Framework for the Precautionary and Inclusive Governance of Food Safety in Europe* Report, co-funded by the EC (Dreyer et al. 2008).

IRGC is an independent, nonprofit science-based think-tank aimed at "providing risk governance policy advice for key decision-makers" (IRGC 2016a). The think-tank has devised the Risk Governance Framework, a comprehensive approach to help tackle deficits in risk-governance structures and processes, including loss of public trust, cost of inefficient regulations, inadequate consideration of risk–loss and risk–benefit trade-offs as well as inequitable distribution of risks and benefits between countries, organizations and social groups (IRGC 2005, 5). The Framework consists of five linked phases: risk pre-assessment, risk appraisal, characterization and evaluation, risk management and risk communication (IRGC 2016b). What is importance to notice is the addition of the "pre-assessment" and "characterization and evaluation" stages to the typical threefold risk-assessment, risk-management and risk-communication model, during which social factors and perceptions are considered alongside scientific evidence in order to determine how an issue is framed by different stakeholders and whether the risk it poses is tolerable or intolerable. The IRGC is very explicit in stressing the salience of public opinion and, it seems, one of its major contributions to models of inclusive risk governance is precisely the incorporation of public opinion in the rigorous deliberation process. It has to be noted, nonetheless, that the central problem of subjectivity and vested interests in scientists and officials is not adequately addressed in the IRGC. In applying the Framework to the case of GMOs, Joyce Tait, director of the Innogen Institute, argues that "ideally, public policy makers and regulators should take the lead in managing the framing of the risks and benefits of new technology to minimize the biases likely to be introduced by both industry and public advocacy groups" (Tait 2012, 11). By suggesting that policymakers and regulators take the lead in the framing of risks and benefits, the problem of uneven public involvement is in essence sustained within the tacitly accepted distinction that some social actors are less prone to COI and ideological motives than others.

The IRGC Framework, as a promising platform for inclusive risk governance policy advice, shares common deliberative premises with the *General Framework for the Precautionary and Inclusive Governance of Food Safety in Europe* (in short, the "General Framework"). The General Framework constitutes a concrete proposal as to how decision-making on food-related issues should be conducted on an EU level and comes as a response to food-related governance challenges that include, among others, the handling of scientific uncertainty, the coordination between assessment and management of food-safety threats, the involvement of diverse social groups in the governance process and the handling of highly controversial food-safety issues (Dreyer et al. 2008, 5). This proposed governance cycle consists of four stages: framing, assessment, evaluation and management. Effective communication and public involvement constitute an integral part of all four phases of the Framework, in this way offering multiple opportunities to stakeholders to share their own knowledge and values and, at the same time, communicate to the wider public information on the process and its results (Dreyer et al. 2008, 33). In the very detailed report, specific policy measures that can stimulate genuine public engagement are suggested. These include the creation of an *Internet Forum* to serve as a web-based platform for involvement and debate on issues of framing, assessment, evaluation and management; and an *Interface Committee* to help structure face-to-face discussions between assessors, managers and stakeholders (Dreyer et al. 2008, 68–78). Overall, the General Framework can be regarded as a viable and concrete policy plan on food safety, one that is guided by the precautionary principle and the idea of inclusive governance "understood as the obligation to ensure early and meaningful involvement of all stakeholders and, in particular civil society" (Dreyer et al. 2008, 34; Jasanoff 1993).

The significance of a consensus-based mediated dialogue process

One of the major arguments developed in this chapter is that scientific disagreement is very often fueled by idiosyncrasy. Lack of consensus can be routinely associated with masked conflicts of interest, ideological differences that use science as a pretext, deeply seated dispositions, emotions of distrust toward the other party, feelings of anger and frustration and so forth. The GM controversy has proved to be a fertile playground for the manifestation of such instances. If the significant role that a situated actor's internal structures and active agency play throughout the course of the debate is seriously taken into account, then it becomes obvious that these elements need to be openly acknowledged, articulated and appraised in an environment that encourages face-to-face interaction and has at its core the human factor. This point is, in fact, explicitly made by Harold Saunders (2001) and the authors of the General Framework, who caution that in cases in which "food safety problems are subject to strongly divergent cultural attitudes, political perspectives, or economic interests, it might be required to […] organize *face-to-face* participatory deliberation processes involving all relevant stakeholders and/or representatives of the wider public" (Dreyer et al. 2008, 64, stress in original). As a response to this challenge, it is argued in this section that formal policy frameworks can be complemented by an informal consensus-based approach that propounds the need for participatory deliberation.

In dealing with the science-intensive dispute of GMOs, therefore, we have to take into account the issues of scientific disagreement, the plurality of involved actors and the role the human dimension plays. We can start developing a model that accommodates all three factors—based on the assumption that disputes containing a scientific component in general, or environmental disputes in particular, can be constructively addressed in an environment of mediated dialogue (Bercovitch 2009; Ozawa 1991, 1996; Ozawa and Susskind 1985; Sjöstedt 1993, 1994, 2009; Susskind 1994; Renevier and Henderson 2002; Zapfel 2002). Mediation is a voluntary process distinguished from negotiation by the inclusion of a commonly accepted *nonpartisan facilitator*, a person or a group of individuals, responsible not only for handling the most mechanical aspects of negotiation—such as scheduling meetings and keeping records—but also for more substantive functions such as ensuring a shared understanding of technical points among all participants, suggesting courses of action for helping resolve contested points and proposing alternative formulations of agreements (Ozawa and Susskind 1985, 32). The facilitator in the suggested GM dialogue will, therefore, be responsible for tending to the formal and substantive aspects

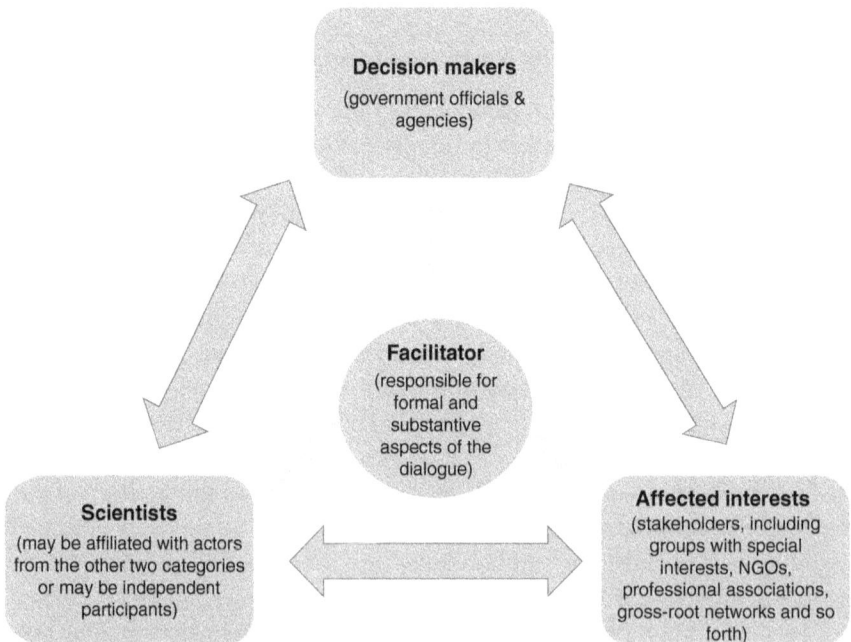

Figure 8.1 The basis of a consensus-based mediated dialogue.

of the process in a way that the involved participants (government agencies and officials, various interest groups and private individuals, along with their respective scientific advisors) find acceptable. The founding block of this procedure, which comes in stark contrast to formal decision-making frameworks, is that all types of participants are brought on a par. No primacy is given to scientists, decision-makers or groups with high economic capital, because the opinions, evidence and objections presented by all of them are attended to by the facilitator, who is placed at the center of the proceedings. The initial understanding of the procedure can be depicted in Figure 8.1.

If the platform of such an initiative is accepted, then we can start sketching the procedure that should be followed. In decision-making processes, science is used in four phases: *agenda-setting, problem formulation, identification of alternatives* and *decision choice* (Ozawa 1991, 80). These four stages are steps that are neither necessarily distinct nor sequential in decision-making, as they may overlap, and the process may in some cases repeatedly vacillate between two or more phases (ibid.). Having said this, it seems to me that these four steps can be constructively reconceptualized as *stages of the consensus-based dialogue process*. It has already been argued, however, that the issues of scientific disagreement and the subjective world of each actor need to be accommodated in the proposed dialogue initiative. As a result, the dialogue process can consist of the following stages: *deciding to engage, mapping problems and relationships, agenda setting, problem formulation, fact-finding, identification of alternatives* and *decision choice*. The first two steps have been adopted from the concept of sustained dialogue—or Public Peace Process—developed by Saunders (2001) as a tool for transforming deeply rooted racial and ethnic conflicts, while the stage of "fact-finding" has been identified by Laurence Susskind (1994) as a core step in environmental negotiation processes. We can now turn to analyze what this initiative entails.

Seven steps toward dispute resolution

Stage one: Deciding to engage   This is the cornerstone of the consensus-based dialogue and entails a set of challenging tasks that need to be addressed, such as: Who will take the initiative? Who has the will and capacities to engage in dialogue? In what space and under what conditions will the dialogue take place? Who can facilitate the decision to make a meeting happen and how? It is of vital importance that this stage is completed, with all the parties being sufficiently pleased with the decided-upon arrangements. We have seen in this book, for instance, how dialogue initiatives have failed because the time frame was too restricted (e.g., Take the Flour Back and Rothamsted Research) or because the overall conditions of the dialogue were deemed to be loaded (e.g., FSA dialogue).

The initiative for dialogue can be taken "from within," that is, from actors already involved in the controversy and who want to reach out to the other side; or "from outside" by a professionally trained third party who wishes to assist constructively in overcoming the tension: such a body could be, for example, the IRGC. Once there is agreement among parties that there is indeed merit in initiating discussions, the participating actors need to be brought together. As already suggested, these should include representatives of the three broad types of stakeholders. Since, however, these three groups do not constitute homogeneous wholes, but simply analytical categories that consist of highly diversified elements, it is imperative that the party that has initiated the procedure invite not only the most vocal and salient members—what we could call macro actors—but also "a randomized or deliberately stratified group of individuals" (Dreyer et al. 2008, 58) so that inclusivity and diversity of representation is achieved. Regarding scientists, particular attention needs to be paid so that a sufficient *representation across disciplines* is ensured. In this way, the "lightning rod" function of the GM controversy and its multilayered complexity are acknowledged by encouraging a dialogue that, in the subsequent stages, fosters alliances across disciplines, especially between social scientists and natural scientists (Zapfel 2002, 148). The list of participants cannot, of course, be compiled within a single process, but may gradually emerge during this phase while discussions between parties unfold. Finally, at this stage, the facilitator (who may be either a single person or a group of co-facilitators) is appointed by consensual decision. While each side may seek for particular qualities in a facilitator, some essential attributes usually include the following: commitment to the overall purpose of bridging the scientific, ideological and framing differences between the stakeholders; realistic expectations for the pace at which consensus may be forged; some experience with related problems and the ability to draw on that experience in this particular case; the ability to help individuals see shared elements in their views and experiences; the ability to help participants organize their thoughts; and the capacity to set agendas that build from previously expressed ideas to help advance and deepen the dialogue process (Saunders 2001, 98–109).

Stage Two: Mapping and Naming Problems and Relationships   The purpose of this phase is to bring to the surface the full range of significant problems and the dynamics of the relationships that cause them: the irritants, dilemmas, misunderstandings and practices that reflect sources of difficulty, conflict or opportunity in the relationships (Saunders 2001, 115). This means that the process of mapping problems and relationships is necessarily wide-ranging and may last through several meetings (Saunders 2001, 111). The facilitator may initiate discussion by asking broad questions, such as "Why do you think GMOs have been banned/allowed in this country?" "How does this affect your own decisions and interests?" "What do you think of the technology behind GMOs?" "Why do you say other people do not share your opinion?" "What kind of agricultural policy would you like?" "What alternative to the present situation would you like to move forward?" All such questions may instigate an interesting and broad discussion, which can help identify the dispute's "important dimensions, the relationships that cause it and the interests affected by it" (Saunders 2001, 61). In this phase of deliberation, each participant's formal affiliation, individual and group interests, and subjective understanding of the situation should come to light and serve as a basis for dialogue (Saunders 2001, 115). This step, in essence, recognizes the political nature of scientific disputes and appears to encourage decision-makers and policy players to state explicitly their own concerns. By increasing the participants' understanding of competing and conflicting interests, the level of discussion is elevated, and the prospects of bridge building are increased (Ozawa 1991, 74). Toward the end of Stage Two, participants should now be able to talk *with* each other instead of *past* each other and simply state their own views, showing readiness to settle down to an in-depth discussion of specific problems, one at a time (Saunders 2001, 119).

Stage Three: Agenda Setting   Building on the conclusions of the previous phase, during Stage Three participants decide which aspects of the GM controversy will be assessed during the process. This decision is primarily contingent on the type of agenda under discussion. Roger Cobb and Charles Elder (1971) define two broad kinds of agendas: the systemic and the institutional. The systemic agenda is more abstract, more general, and wider in scope and refers to "a general set of political controversies [...] falling within the range of legitimate concerns meriting the attention of the polity" (Cobb and Elder 1971, 905–6). On the other hand, the institutional agenda denotes "a set of concrete items scheduled for active and serious consideration by a particular institutional decision-making body" (Cobb and Elder 1971, 906). Depending, therefore, on the overarching nature of the agenda, a corresponding classification of topics needs to be made. If, for instance, it has been agreed by the participants that the agenda will be of a more general and broad nature (i.e., systemic), the items to be discussed will also reflect this more abstract orientation; they can include, for instance: the role GMOs can play in a country's agricultural policy, the funds that should be allocated to innovations in agbiotech, how the challenges of food security can be addressed, the relations between GMOs, conventional and organic farming and so forth. If, on the other hand, the mediated dialogue aims at clarifying concrete items in need of active and serious consideration, then the issues will, of course, be more targeted. Such issues may include: the health effects that a particular type of GM maize (MON810, for example) allegedly has on rodents, the adverse effects that GM *Bt* maize has on monarch butterflies, the legal protection of farmers whose crops unintentionally contain GM material as a result of gene flow and so forth. The list of such items on the agenda is contingent on various factors, such as the time and funding available for the dialogue process, the needs of policy reformulation, the interests of the actors, the degree of expertise, the status of the actors' conjuncturally specific knowledge and more. In order to increase the chances of the dialogue being effective, apart from classifying the items on the agenda so that they comprise a more or less homogeneous set in terms of levels of ontological scale and abstraction, it is equally important to classify them according to the nature of the problem. While no case can be matched to a single discipline, and an interdisciplinary panel is required at all times, undoubtedly certain issues demand more expertise than others in a particular field. The question of feeding trials on rodents, for instance, demands a significant presence of qualified toxicologists; the effect on monarch butterflies also requires environmental scientists; the concerns about corporate control of the food chain require legal expertise and so forth. This should be taken into account when the agenda is set so that central issues to the GM dispute are not conflated or discussed *en bloc* but are allocated to panels where scientists have the appropriate expertise.

Stage Four: Problem Formulation   At this stage, the agenda that has been consensually agreed upon is operationalized. The issues identified as topics of discord that need to be addressed during the dialogue process will now become specific hypotheses in need of specific analysis and assessment. As in all stages, there has to be an agreement in how problems are formulated, and it is one of the facilitator's main tasks to ensure that the comments and contributions of all stakeholders are accommodated during this phase. According to the agreed structure and the magnitude of the dialogue, if it has been decided that multiple problems will be allocated to the equal number of panels (each consisting of affected interests, decision-makers and scientists), then the allocation should be done on a thematic basis, which will be reflected in each panel's synthesis of appropriate scientific expertise. Having said this, it is usually more appropriate to address issues requiring natural scientists first, so as to help clarify the basic considerations of the technology and tackle social problems at a later stage, when the risks and benefits of the technology have been acknowledged (Zapfel 2002, 133).

Stage Five: Fact-finding   This stage aims at bringing parties together in an effort to "establish a baseline of facts on which there is agreement and to clarify the scope and nature of differences in the understanding of the problem" (Susskind 1994, 62). On the issues identified in the previous stage, stakeholders can now establish whether it is true that, for instance, *Bt* GM maize actually impacts the mortality rate of monarch butterflies; or better, to what extent this is true, how high the rate is, under what environmental conditions and so on. The end of this stage should find participants with a very clear view on the issues they agree on—which may be particular sub-hypotheses of agenda items—and the issues they disagree on and the reasons why the disagreement persists.

Stage Six: Identification of Alternatives   After the facts have been established and the remaining differences have been explicitly stated and justified, alternative solutions need to be proposed. By acknowledging the baseline of consensual facts and the topics of discord, stakeholders need to propose viable alternatives so that the items placed on the agenda are no longer contentious. How can, for example, public funds be allocated to agbiotech innovations in a way that both stimulates growth but also gains public approval? In what ways can farmers be legally protected from unintentional gene transfer and, at the same time, how can the implications of the GM seed patent be respected? What measures need to be taken so that monarch butterflies and other Lepidoptera are protected from the suspected harm caused by GM plants? This stage is completed when alternative propositions for addressing the agenda items are placed on the table.

The role of the facilitator in Stages Four, Five and Six   The three previous stages are of particular importance in science-intensive debates such as the one under discussion. This is so because it is during these three stages that information, facts and research models are discussed, accepted, disputed and reformulated. The role of the mediator during this time is not only to ensure that information and opinions are clearly circulated among the participants, but also to tend to the crucial matter of scientific discord that seems to torment the GM debate. Three types of techniques can be routinely employed during the three stages of problem formulation, fact-finding and identification of alternatives: *information sharing, joint fact-finding* and *collaborative model building* (Ozawa and Susskind 1985, 32).

Information sharing is presumed to be socialized in the scientific community. This means that, going back to the Mertonian analysis of scientific norms discussed in Chapter 2, scientists are ideally inclined to share newfound knowledge in their quest for "scientific truths." While, as already discussed, this norm is rarely instantiated in the GM controversy because personal animosity and ideological differences often get in the way, in a mediated environment, participants "are encouraged to see information as a means of opening up new possibilities for dealing with differences" (ibid.). Scientists who choose not to divulge information—which was the case with TestBiotech and the anonymous toxicologist—run the serious risk of losing credibility as mediation unfolds in front of other members of the scientific community, government officials and various affected interests. This practice also safeguards against the tendency of companies and public-interest advocacy groups who choose to "quote selectively from evidence that supports their case" (Tait 2012, 11), since doing so in this environment will hurt both public and governmental confidence in their impartiality. Finally, since in the suggested mediated environment decision-makers do not hold any authority but are situated as mere participants in the debate, they can demand and receive information in whatever form they feel best exhibits the conflicting claims of affected interests. In this way, they acquire a deeper understanding of critical issues, which can foster more-informed policy decisions (Ozawa and Susskind 1985, 32–33).

Second, mediation in the GM dispute can—and should—encourage *joint fact-finding* and *model building*. During this process, scientists can jointly frame the research questions, build theoretical models, specify the method of inquiry, appoint researchers, assess their work and voice their concerns at every appropriate step. If this initiative on the part of the facilitator is accepted, then a significant step toward a common course is made since, if parties make such central decisions together, then the possibilities of rejecting the final results significantly diminish. Furthermore, such an exercise will also benefit decision-makers and affected interests, as their own understanding of the issues will most likely advance (Ozawa and Susskind 1985, 33).

By encouraging collaboration among participants, the facilitator is trying to address the roots of scientific disagreement discussed in the previous section: miscommunication, differences in the design of inquiries, error in the inquiry and differences in the interpretation of findings. The endeavor to limit such differences also achieves a considerable blow to the practice of concealing interests and ideological schemata behind scientific findings. In this way, the facilitator acts as a "guardian of the process" by intervening to correct miscommunications, clarify ambiguous messages, challenge deceptive communications and point out when differences in interpretations have arisen (Ozawa and Susskind 1985, 35). While each actor's interests, motives, dispositions and subjective reading of the situation will always be significant, within the mediated environment of cooperation their role becomes more controlled and distinct.

Stage seven: Decision choice   The mediated dialogue ends when the alternatives presented have been discussed and, among an array of formulations and corresponding alternative actions, the participants have reached an agreement as to the conclusions the process has yielded. This may include a commonly accepted agreement on an agenda issue or may acknowledge headway made on certain aspects, while other particularities may still cause discord. The decision choice can serve as a vehicle for policy change or as a starting point for a new cycle of mediated dialogue.

Mediated dialogue: Backward and forward

The first major argument of this chapter was that the GM controversy should not be construed as a dualistic case of scientific progress against political interventionism, since there is no scientific consensus on the safety and benefits of GMOs and the political field very often endorses ventures in (agri)biotechnology. Instead, it has been suggested that the GM controversy is an embracing ecological dispute with its distinct characteristics and plurality of concerned stakeholders.

The second major component of Chapter 8 has been the exploration of the root causes of scientific disagreement and the complex ways by which objective criteria and subjective *lifeworlds* intermesh. By drawing on various examples from the GM conjuncture, it was demonstrated that advocates and skeptics of the technology, alike, regardless of their academic qualifications and socioeconomic status, can be equally emotional, committed to ideological motifs and susceptible to conflicts of interest.

The two argumentative threads were brought closer together in the final section of the chapter, where the merit of public involvement was discussed and it was argued that the crucial role that personal idiosyncrasies play in the development of the GM debate can be fruitfully accommodated in an informal and mediated environment of dialogue. This suggestion comes as an alternative offer to voices that demand dialogue of greater length and more participants. Instead of focusing on a quantitative expansion of current initiatives, it has been argued that a change in perspectives may prove to be even more effective in alleviating public distrust, scientific disagreement and idio-syncratic decision-making. By critically synthesizing models of participation from the multi-disciplinary field of conflict resolution, a seven-stage model of consensual-based mediated dialogue was offered. These seven stages, it should be noted, are neither strictly sequential nor obligatory. The dialogue may often vacillate between Stages Two and Three or Three and Four, or may terminate before the completion of Stage Five, during Stage Six or even Stage Two. The openness of the participants, their sincere intentions of understand-ing the other side's view and worries, and the facilitator's skills, all play a role in how the dialogue will eventually evolve. What is more, the suggested frame-work is quite flexible as to its scope and complexity. It can be used for the discussion of issues on systemic or institutional agendas, for deliberation on a particular topic or a cluster of topics, or for mediated dialogues taking place on national or international levels. Despite the degrees of freedom this model allows regarding its magnitude, it is limited in its ability to accommodate a large number of participants, as this would jeopardize the effectiveness of its mediated, face-to-face character. The suggested informal dialogue, together with the formal national and international decision-making processes, could very well work in tandem as parts of a holistic risk-governance framework whereby each part informs the other about headway that has been made on particular agenda items.

Finally, it needs to be repeated once again that proposing a fully fledged risk-governance framework certainly lies well beyond the scope of this book. Having said that, a schematic presentation of a sustained dialogue model was sketched in the concluding pages for reasons of argumentative clarity and completion. It is my firm belief that its tentative nature will greatly benefit from future research which will embed it across critical contours of the social order, such as: the public attitude toward science and technology; the role of the media and popular representations; and the ways in which these can shape popular perceptions, the concept of the public sphere, and the lessons that can be drawn from other practical experiments in organizing deliberative democracy.

Chapter 9

CONCLUSION

By now, I hope to have offered a different, more profound understanding of the Rothamsted GM wheat trials, in particular, and the GM controversy in general. By drawing on key literature in STS and notable contributions to contemporary sociological theory and conflict resolution, I hope to have demonstrated the central role social theory can play in our understanding of current conjunctures in the field of science and technology. This understanding is unsatisfactory, however, if it is not accompanied by courses of action aimed at a more inclusive and democratic social world. This book has suggested one way, among many, that the GM controversy can be approached theoretically and reframed practically.

The broad theoretical framework was built piece by piece in the first five chapters of the book. In Chapter 1, the need for a theoretical framework was recognized, as the newsworthy events of the Rothamsted field trials brought to the surface recurring themes of the broader GM debate. These actions were not limited to the rather expected instances of public protest, but also broached crucial issues such as the scientific disagreement on the safety and benefits of GMOs, concerns about corporate control of the food chain, worries that propagation of GM technology will come at the expense of other agricultural trajectories and more. These issues were also present in the public inquiries that the UK government launched throughout the past two decades. Nonetheless, regardless of the reservations expressed by scientists, professional associations and bodies of the civil society, in 2015 the UK government decided to allow cultivation in English soil, while Northern Ireland, Scotland and Wales all opted-out, citing scientific uncertainty and public apprehension regarding the technology. At the very least, what all this tells us is that the GM debate is not comprised of a litany of free-floating events, but is a conjuncture embedded in the broader social, political, economic and cultural environments. It follows then that this is not a scientific issue, but a topic which, like a lightning rod, attracts much broader concerns. Such concerns are deliberated across a diverse and heterogeneous network of position-practice relations, which includes, for instance: the protesters at the field of Rothamsted who wanted to safeguard nature from the adventitious flow of genetic material; the scientists in front of the House of Commons Science and Technology Committee expressing their doubts about the safety and usefulness of GMOs; the agbiotech corporations investing millions in R&D in order to eventually commercialize crops with unique traits; and the technocrats in Brussels devising draft after draft of a framework that could accepted by the majority of member states, just to mention a few. What distinguishes these actors is not merely the positions they occupy in various structures with their respective institutional and normative arrangements, but also their own subjective worlds, how they interpret the current situation, how they construe other people's actions, their own dispositions, their more specific knowledge of a situation, their interests and motives, the broader ideological elements they espouse and more.

Having established the main points that the analysis of the research question should address, Chapter 2 started with an overview of the broad avenues that theorize science, technology and society relations: namely, technological determinism, social constructivism, Misa's "meso-level approach," Hughes's "technological momentum" and Jasanoff's idiom of "co-production." Chapter 3 analyzed the most influential approaches in STS and highlighted each offering's heuristic value: its ontological claims, its methodological premises and how these two help the researcher grasp the dynamics of social reality. On these grounds, each approach's offering that was found ontologically sound and methodologically appropriate for the research project at hand was taken on board as part of the process of critically synthesizing the most notable contributions to the field. One main issue that emerged from the review is the imbalance that exists in theorizing human/nonhuman relations and the problems that accompany the three main alternatives. More precisely, variations of technological determinism give primacy to the technoscientific (nonhuman) aspect and assert that social change is merely a corollary of it; social constructivism claims the exact opposite by bestowing quasi-Promethean powers on humans while post-humanist approaches deconstruct the boundaries between the two with the usage of a hybrid language.

Chapter 4 started with the work of Ted Benton and his concept of ontological naturalism, where humans and nonhumans are placed on a continuum, and their shared characteristics are explored. The chapter also familiarized the reader with the main principles of Nicos Mouzelis' post-Marxist Technology-Appropriation-Ideology scheme and Rob Stones's Strong Structuration Theory (SST). Chapter 5 revisited the contributions of the three key sociological thinkers from a more reflective standpoint, with the aim of forging ontological naturalism, TAI and SST into a holistic framework appropriate for the study of agricultural biotechnology. To reiterate, operating at different levels of ontological scale and abstraction, Mouzelis's framework, after some necessary modifications, invites the researcher to examine the field of agbiotech along its three constitutive dimensions: the Technological (the technology of GMOs; where, how and under what arrangements it is produced), the Appropriative (the groups of people that try to control Technology and the means they use) and the Ideological (the discursive schemata used by stakeholders in order to justify their actions). SST, on the other hand, is more suitable for assessing how actions unfold and relationships emerge by coherently conceptualizing the role that the external environment of action, an actor's internal structures (typification of things, worldviews, habits of speech and so forth, and conjuncturally specific knowledge of the situation) and their active agency (how interests and needs are prioritized, the horizon of action, the unconscious and more) all play a role in how social order is produced, reproduced and altered.

The heuristic value of the suggested framework was explored in chapters 6 and 7. In Chapter 6 the Rothamsted GM wheat trials were examined along their Technological and Appropriative dimensions, where the analysis focused not only on the particularities of the specific experiment (its scientific significance, its social and cultural connotations, the involvement of specific actors and their idiosyncratic behavior) but gradually led to the wider issues underlying the GM controversy in the UK and the EU and the adequacy of the current regulatory framework. The significance of *Nature* and the precautionary principle were discussed in Chapter 7 as contentious ideological constructs that bring to the surface the wealth of broader differences GM advocates and skeptics have on the notions of progress, democracy and welfare. The sectarian character that the GM controversy often takes was further demonstrated in Chapter 8, where the commonly held framing of the debate as a case of science versus politics was discussed. The lack of consensus within the scientific community on the benefits and risks of GMOs was demonstrated and, in line with the view that scientists are not at all immune to conflicts of interest and ideological perceptions, the neat compartmentalization of the two spheres was blemished. It was argued, instead, that the GM controversy should be more accurately reframed as an embracing ecological dispute of low intensity. This proposition has not only theoretical merits, as it sheds light to the various contours of the conjuncture that often go unnoticed, but it has practical implications as to how it can be resolved. Gathering pertinent elements from various models of conflict resolution from the homonymous multi-disciplinary field, an informal model of a consensus-based dialogue was schematically introduced in order to complement formal frameworks of risk governance.

Things change very fast in the field of agricultural biotechnology. The more recent development that will likely add more fuel to the intense debate, if it has not already, is the struggle to regulate New Breeding Techniques (NBTs). By the first quarter of 2016, the EC is expected to have reached an opinion as to whether NBTs should fall under EU GMO legislation or not (Michalopoulos 2016). It should not come as a surprise by now that scientific, public and regulatory opinion is divided on this issue as well. NBT techniques, such as cisgenesis, intragenesis, agro-infiltration, synthetic genomics and more (European Commission 2015b) are considered by some as different from GM technology as, in the case of NBTs, genes are transferred from related organisms or even the same organism to the resulting plant and, as a result, no foreign DNA is present (Michalopoulos 2015). Skeptics, on the other hand, claim that while this is true, the same laboratory methods used in genetic engineering are also used in cisgenesis and, as a result, this "can have the same disruptive effects as transgenes on the genome, gene expression, and a range of processes operating at the level of cells, tissues and the whole organism" posing "most of the same risks to health and the environment as transgenic GMOs" (Fagan, Antoniou and Robinson 2014, 52). It is very likely that NBTs will signify the second round of debate that was initiated with GMOs (Holland 2016).

If the framework suggested in this book has helped offer a more robust understanding of the GM debate, then it can also serve as a flexible approach to the reader who wants to acquire a critical understanding of various issues in agbiotech, such as NBTs, and to the researcher who wishes to examine these topics with greater rigor. Regardless of the purpose and the methods used, it is my sincere conviction that policy issues, especially those that entail risk and scientific uncertainty unfolding at the frontiers of science and technology, should be approached with humility, openness and commitment to sincere dialogue. This book has been one such offer toward this end.

NOTES

Chapter 1

1 For details and examples of such incidents, see Levidow and Carr (2010, esp. chaps. 3 and 7).

2 When the two categories of "GM advocates" and "GM skeptics" are used throughout this book it is done purely for purposes of analytical convenience and narrative parsimony. The term GM advocates refers to those individual and collective actors who have overtly expressed an opinion, which construes the potential benefits derived from GMOs as more salient than the probable associated risks; conversely, GM skeptics refers to those actors who regard the potential risks as greater than the probable benefits. Having said that, this crude categorization certainly respects the following facts: first, there is a constant flux in the construct, shape and change of attitudes; second, there are certainly variations within each group, ranging from more malleable viewpoints to more intractable ones; third, there is no a priori implication as to why an actor follows a more embracing or cautious approach to the issue; fourth, there may be an overlap on certain aspects of opinions among members of the groups; fifth, many actors do not fit in either of these two categories, as they are either undecided or their opinions have not been explicitly stated and consequently are unknown to the external researcher. As one of the book's main arguments is that the GM controversy is a rich and multilayered issue and should not be seen as a binary choice, the use of the two categories is employed as rarely as possible and only on instances when an alternative narrative framing would simply be too cumbersome.

3 The UK is the only EU country still maintaining a full-time CSA post in its government. The other two countries that also had established the position—the Czech Republic and Ireland—have dismissed the specific post (Nelsen 2014).

4 In their letter to the PM, cosigned by Sir Mark Walport (the government's CSA), the CST invited the PM to visit Rothamsted Research and provide further briefing on the huge potential of GM technology. More precisely, the cochairs informed the PM that "If you have an opportunity in the future, the researchers at Rothamsted would, of course, be delighted to show you their worldleading and historic research" (Council for Science and Technology 2014, 4). The GM wheat trials are also mentioned in the *GM Science Update* (Baulcombe et al. 2014, 18), the written evidence submitted by DEFRA and BIS to the STC Inquiry (DEFRA and BIS 2014, para. 15, 21–23) and also the written evidence submitted by Dr. Charles Clutterbuck (Clutterbuck 2014, 2), Sense about Science (Sense about Science 2014, Appendix 1) and The Science Council (Science Council 2014, para. 4.9) to the same inquiry.

The field trials are also promoted as a "good example of ground-breaking UK agricultural biotechnology research" in the *Going for Growth* report (Agricultural Biotech Council 2012a, 7).

5   See, for example, the excellent works of Britt and Lappé (2002), Charles (2001), Falkner (2007), Higgins and Lawrence (2009), Horlick-Jones et al. (2007), Howlett and Laycock (2012), Kinchy (2012), Kleinman (2003), Levidow and Carr (2010), Levidow and Murphy (2006), Schurman and Munro (2010), Toke (2004).

Chapter 2

1   See, for example, the documentary *The Machine that Changed the World* (1992) about the advent of personal computing and the social transformations this technology brought about (Linde 1992).

2   Whether Marx was or was not a technological determinist is a matter of a lengthy and sometimes bitter debate among Marxist commentators (Shaw 1979, 155). For one thing, MacKenzie (1984) and Bimber (1990) argue that with closer reading it becomes apparent that despite these overly simplistic generalizations found in the "hand-mill" passage, Marx was clearly not an advocate of technological determinism. "[T]he thesis that Marx was a technological determinist in any strong sense [is] extremely difficult to sustain, at least without invoking a peculiar and marked inconsistency between his general statements and particular analyses" (MacKenzie 1984, 480). Misa also agrees with this viewpoint by arguing that: "Marx the macro analyst penned statements (such as the above [i.e., the "hand-mill" and "railway" quotations]) that affirm technological determinism, but Marx the micro analyst rejected technological determinism" (Misa 1994, 123).

3   To be sure, Cohen offers an extensive and very detailed account of the *primacy thesis* in Marx (Cohen 1978, 134–74). "The primacy thesis is that *the nature of a set of production relations is explained by the level of development of the productive forces embraced by it* (to a far greater extent than vice versa)" (Cohen 1978, 134, stress in original). It is obvious that, since technologies can be found in productive forces, their significance is more or less automatically upgraded with the acceptance of the primacy thesis. Nonetheless, Cohen's discussion clearly eschews direct references to technological determinism, with the notable exception of the distinction mentioned above.

4   Smith and Marx call this twofold distinction as the "hard view" of technological determinism. This kind of "hard determinism" is also an example of what Bimber (1990) calls "logical sequence accounts," which he identifies in the works of Cohen, Miller and Heilbroner. I think Bimber's (1990) analysis of technological determinism is rather misleading. If Heilbroner clearly states that his study "will enable us to test the empirical content—or at least see if there *is* an empirical content—in the idea of technological determinism. I do not think it will come as a surprise if I announce now that we will find *some* content, and a great deal of missing evidence, in our investigation" (Heilbroner 1994, 55); and if Smith and Marx in the preface of that article state that Heilbroner "[u]ltimately views technology as a strong 'mediating factor' rather than as the determining influence on history—a point he reiterates and expands upon in the retrospective essay that follows this one" (Heilbroner 1994, 53), then one is left to wonder how and why Heilbroner qualifies as a technological determinist.

5   Without it belonging to the school of technological determinism, I think it is worth mentioning a loosely related concept that Benton (1994, 31–38) calls "technological optimism" and Kleinman (2005, 3) "technological progressivism." This view, which can be traced back to the Enlightenment, advocates a linear, evolutionary view of history by which science and technology are indiscriminately endowed with positive attributes and construed as "stepping stones of human development" (Wyatt 2008, 169). For this point of view, technology is seen as asocial and self-propelling (Kleinman 2005, 123). Technological optimism/progressivism does not espouse the belief that the impact of technology on society is irreversible. As Benton (1994, 36) mentions, "[t]he technological determinist picture is false, but it is made *plausible* by an inscrutable concentration of power over science and technology in both advanced capitalist and state socialist societies" (stress in original). In other words, the inevitability of a trajectory is advocated by those who are in power and use this facet of technological determinism as an ideological mantle in order to avoid discussions regarding different paths that can be followed. To give an example, in their study of Monsanto's discursive products in its battle to gain public support for the development of biotechnology, Kleinman and Kloppenburg (1991) demonstrate how slogans and images are used in order to reinforce the tenet of technological optimism/progressivism.

6   For an actor-essentialist view of collective action, see Olson (1998).

7   As is demonstrated in the next chapter, there are crucial differences among some of these theorists.

8   This is the exact opposite of what Kleinman (2005) calls "scientism." This belief, which powerful companies readily espouse, ascribes "cognitive superiority of facts over values [and] leads to the conclusion that only trained scientists—*experts at unearthing facts*—can appropriately participate in decision-making on technical matters" (5, stress in original).

9   "There is a long history of technological prediction, some of it ludicrous and some not. What is interesting is that the development of technical progress has always seemed *intrinsically* predictable. This does not mean that we can lay down future timetables of technical discovery, nor does it rule out the possibility of surprises. Yet I venture to state that many scientists would be willing to make *general* predictions as to the nature of technological capability 25 or even 50 years ahead. This, too, suggests that technology follows a developmental sequence rather than arriving in a more chancy fashion […] In the future as in the past, the development of the technology of production seems bounded by the constraints of knowledge and capability and thus, in principle at least, open to prediction as a determinable force of the historical process" (Heilbroner 1967, 338–40, stress in original).

10 In Collins's words: "We must treat our perceptions of the world, for the purposes of this exercise, like 'pictures in the fire'. If the world must be introduced then it should play no more role than the fire in which the pictures are seen. Or better—think of one of those pictures which one constructs by joining numbered dots with pencil lines. Now imagine the world consisting of a large sheet covered in almost infinitesimally small dots. The world is there in the form of the paper but mankind may put the numbers wherever he wishes and in this way can produce any picture" (Collins 1992, 16).

Chapter 3

1   For a detailed analysis see Merton (1942). It seems to me that, instead of criticizing Merton on the grounds that the ethical norms of science that he proposes (namely, universalism, communism, disinterestedness, organized skepticism) are no longer pertinent to contemporary scientific practice, one should construe them as Weberian ideal types in order to see whether norms of another field have "colonized" science. For example, if the value of "communism," which means that "the substantive findings of science are a product of social collaboration and are assigned to the community," is in direct contrast with the practice of "patents," and cannot be readily found in contemporary capitalist societies, this is so because the norms of the economic sphere that urge for competition and maximization of profit have permeated the scientific field. For further analysis see Benton and Craib (2001).

2   Other authors who have contributed to the birth of ANT include Madeleine Akrich, Geof Bowker, Alberto Cambrosio, Cecile Medeal, Arie Rip, Susan Leigh Star and Antoine Hennion (Law 1992, 379–80).

Chapter 4

1   See Archer (1996), Giddens (1993), Ian Goug and Gunnar Olofson (2000), Hall (1995, 1998), Layder (2006), Sibeon (2004), Stones (2005, 2008), Rudebeck, Törnquist and Rojas (1998).

2   See the favorable reviews by Scott (1996, 731–32), I. J. Cohen (1996, 597) and Urry (1992, 692).

3   Engels himself fiercely rejected any reductionist interpretation of the base/superstructure image in which the economy would be construed as the cause and the superstructure as merely the effects. In his letter to J. Bloch (September 21–22, 1890), Engels writes: "Neither Marx nor I have ever asserted more than this [i.e., the idea that the superstructure can produce effects and react upon the base]. Hence if somebody twists this into saying that the economic factor is the *only* determining one, he transforms that proposition into a meaningless, abstract, absurd phrase" (qtd. in Bottomore 1991, 47; emphasis in the original).

4   See, for example, Bimber (1990) and Young (1976).

5   Giddens asserts that the duality of structure is based on "the essential recursiveness of social life, as constituted in social practices: structure is both medium and outcome of the reproduction of practices. Structure enters simultaneously into the constitution of the agent and social practices, and 'exists' in the generating moment of this constitution" (1979, 5). It is worth mentioning that, while Mouzelis sees the notion of duality of structure as profoundly flawed, as it only tells us half the story of the agency-structure relation (i.e., the possibility of dualism is not taken into any consideration) (1994, 25–47; 1995, 117–26), Stones, on the other hand, who finds Mouzelis' objections to a great extent justified, finds it more constructive to construe such criticisms as "insightful conceptual refinements" rather than theoretical pretexts in order to abandon Structuration Theory in its entirety (2005, 55–58).

6   Stones treats social and material aspects of the environment of action as two distinguishable but intertwined elements. He clearly states: "My own inclination when dealing with questions about social relations is to treat material and social aspects of the external context as part and parcel of each other—as mutual constituents of the external structures and structuration processes—unless there is an overriding reason for not doing so" (Stones 2005, 199).

Chapter 5

1 Consider, for example, the case of *Monsanto versus Schmeiser*. In 2004, the Canadian Supreme Court ruled in favor of Monsanto Canada against Persy Schmeiser, a farmer from the province of Saskatchewan on the grounds that "knowingly planting seeds that contain patented genes–even if those genes arrived in a field through pollen drift, wind-borne seeds, or other accidental means–constitutes patent infringement" (Kinchy 2012, 102).

2 Stones and Greenhalgh bring human agents and technology closer by sustaining ANT's "actant" terminology and offering a conceptualization of the two under the hybrid umbrella term. While the reservations they express toward the ANT neologism are in perfect agreement with the ones expressed in Chapter 3, I still present here a slightly modified version of SST that keeps the two analytically distinct. This is done in order to avoid even further theoretical discussions that may lead to unnecessary confusion for the reader.

3 The above version of SST can be used to assess interactions with more basic forms of life (or lower-end mechanisms). If, for example, the living organism refers to a gene sequence, then scientist's conjuncturally specific knowledge could be about the acidity of the environment that the genes require to survive, their molecular structure and function, how they tend to affect certain traits, their distinctive characteristics and so forth.

4 This idea is closely related, but not identical, to Archer's (1996) "downward conflationist theories."

Chapter 6

1 At the time of writing, December 2015, the Rothamsted Research Board of Directors numbered ten and was chaired by Sir John Beddington FRS.

2 At the time of writing, December 2015, the Institute director was Achim Dobermann.

3 IDna Genetics provided Rothamsted with the transgene copy number and zygosity testing (Rothamsted Research 2011a, 14).

4 "DNA synthesis was provided by GenScript Inc. Chemically synthesised gene sequences that had been codon-optimised for wheat encoding plastid-targeted enzymes (E)-β-farnesene synthase and farnesyl diphosphate synthase were assembled by GenScript Inc. NJ, USA and introduced into plant cells on complete binary plasmids by biolistic transformation" (Rothamsted Research 2011a, 4, 14). http://archive.defra.gov.uk/environment/quality/gm/regulation/documents/11-r8-01-app-a.pdf.]

5 In the article in *The Independent*, for example, Pickett claimed Pusztai and Ewen changed the rats' diet from raw potatoes to boiled. "I had seen the data and how inadequate it was in terms of tackling the question in hand. I was very critical of the work because it is a shambles really. Rats don't eat raw potatoes very well and half way through they realised this and decided to boil the potatoes" (qtd. in Connor 1999). That is not true, as the rats were consistently fed on diets either containing raw potatoes or boiled potatoes, but never both. No change of diet took place part way through the experiment (Brian 2012).

6 While vice-president of the Royal Society in 1998, Lachmann chaired an expert group that produced the Society's first report on GM crops, entitled *Genetically Modified Plants for Food Use*. The report concluded that GMOs have "the potential to offer real benefits in agricultural practice, food quality, nutrition and health," but cautioned that "there are, however, uncertainties about several aspects of GMOs" (The Royal Society 1998, para. 9). The Royal Society claimed the report was so influential that was used as a "source document" by the UK government (Flynn and Sean 1999).

7   A more detailed account of the genetic modification carried out by Rothamsted reads as follows: "Transgenic wheat plants were produced using standard protocols by microprojectile bombardment. The gene(s) of interest were maintained on separate plasmid vectors each containing a bar gene selectable marker cassette and bombarded on gold particles into scutella of immature zygotic embryos. Whole plants were regenerated and selected from somatic embryos induced in tissue culture. The genes of interest were carried on a binary vector pBract309 (www.bract.org). This was prepared in E. coli Invitrogen DH5α sub-cloning-efficiency competent cells (Genotype F- φ80lacZΔM15 Δ(lacZYA-argF)U169 recA1 endA1 hsdR17(rk-, mk+) phoAsupE44 thi-1 gyrA96 relA1 λ-). Plasmids were purified using a Qiagen plasmid purification Midi kit resulting in the following concentrations: pBract309EBFS+T (clone 6) 1.5mg/ml and pBract309FPPS+T (clone 1) 0.7 mg/ml. Chemically synthesised gene sequences that had been codon-optimised for wheat encoding plastid-targeted enzymes (E)-β-farnesene synthase and farnesyl diphosphate synthase were assembled by GenScript Inc. NJ, USA and introduced into plant cells on complete binary plasmids by biolistic transformation (see Tables and note below). Genes encoding (E)-β-farnesene synthase and farnesyl diphosphate synthase both possessed a wheat chloroplast transit sequence from the small subunit of RubisCo, previously validated to correctly target the proteins to wheat plastids (Primavesi et al 2008). The nucleotide sequences of these genes are synthetic and chimaeric and not found naturally. However, the enzyme encoded by the EBFS cassette is similar to that found in peppermint (Mentha × piperita) and the enzyme encoded by the FPPS cassette has most similarity to that from cow (Bos taurus) but is generally ubiquitous and occurs in most organisms. Both plasmids carry right and left T-DNA border sequences, origins of replication and bacterial selectable marker genes necessary for maintenance in E.coli and Agrobacterium" (Rothamsted Research, 2011a, 3–4).

8   For more details regarding the controversy surrounding the kanamycin resistance gene, introduced by Calgene in 1989, see Rowell (2003, 124–30).

9   During the time of the field trials, the opt-out measure was not a part of the EC directive, and decisions reached at the EU level applied to all member states.

10 The EC's proposal that member states can decide whether GMOs would be allowed in their territories for use in food and animal feed was eventually rejected in October 2015 by a huge majority (557 to 75). The main reason behind the decision was a matter of principle: that is, allowing member states to ban import of GMO products would go against the EU's single-market principle.

11 During the field trials, the EU authorization process followed a different route once the standing committee had either voted "No" or "No opinion." If the Standing Committee on the Food Chain did not agree with the Commission's draft, or if a decision with qualified majority could not be reached, the EC took its position to the Council of Ministers and informed the European Parliament. The Council of Ministers had 90 days for a decision to approve or to reject the draft with a qualified majority. If the Council rejected the Commission's draft, the Commission had to revise its draft. To reach a qualified majority, 232 out of 321 votes were needed. Additionally, a qualified majority meant that at least 62 percent of the EU population was represented. Authorizations were valid for 10 years and subject to further renewal (European Commission 2012a and 2012d; GMO Compass 2006). Furthermore, member states could provisionally prohibit or restrict the use of a GMO on their territory as long as they had new evidence that the specific GMO posed a risk to human

health or to the environment. Despite adopting this "safeguard clause," no member state had ever put forward new evidence (European Commission 2015).

12   At the time of writing—late 2015—the number of directorates is four, as the Science Strategy and Coordination directorate has been axed (EFSA 2015).

13   "EuropaBio is the voice of the European biotech industry. [Its] membership includes a wide range of corporate members and industry associations involved in biotechnology throughout Europe. EuropaBio has 55 corporate and 15 associate members and BIO Regions and 17 national biotechnology associations—representing some 1800 small and medium sized enterprises across Europe" (EuropaBio 2013b). Some of the most prominent members of EuropaBio are BASF, Bayer, Dow AgroSciences, DuPont, GlaxoSmithKline, Monsanto, Novartis, Pfizer and Syngenta (EuropaBio 2013b) EuropaBio represents the three main segments of biotechnology: Healthcare (Red Biotech), Industrial (White Biotech) and Agri-Food (Green Biotech) (EuropaBio 2013a).

14   The idea of "revolving doors" refers to instances in which public officials move to industry jobs or vice versa (Corporate Europe Observatory and Earth Open Source 2012, 2).

15   John Dalli was European Commissioner for Health and Consumer Policy between 2010 and 2012. Commissioner Dalli, who had expressed his outward support of GMOs, was forced to resign on October 16, 2012, when the EU's anti-fraud office, OLAF, linked him to Maltese entrepreneur Silvio Zammit. Zammit was alleged to have asked snuff giant Swedish Match for $77 million in return for persuading Dalli to change the EU's draft tobacco directive (Euractiv 2013a). Following Dalli's resignation, the Commission decided to freeze requests that were in the pipeline to authorize more than 20 GM seeds for cultivation. The biotech industry commented that such delays threaten Europe's food supplies and economic competitiveness, while Mute Schimpf, food campaigner for Friends of the Earth Europe, said: "For us it's a responsibility and a task for a consumer commissioner to listen to the needs and the wishes of the consumers instead of following a handful of biotech companies' interests" (Euractiv 2012d).

16   In a press release issued by EFSA it was stated: "Upon request of the European Food Safety Authority (EFSA), Diána Bánáti has resigned on 8 May as member and Chair of the Management Board, effective immediately. She has decided to take up a professional position at the International Life Sciences Institute (ILSI) which is not compatible with her role as member and Chair of the EFSA Management Board. The Code of Conduct adopted by the EFSA Management Board obliges all members to consider possible public perception, in all facets of their professional and private life, in particular with regard to any activities which could raise doubts about their independence, even with respect to potential conflicts of interest. Board members shall not hold positions or interests that are considered incompatible with their role as a Board member and the role of the Board itself" (EFSA 2012a).

17   According to EFSA's RAE, genetically engineered plants are not seen as technically derived new organisms but as similar (comparable) to conventionally bred plants. Consequently, EFSA does not require comprehensive investigations of the plants per se, although genetically engineered plants are known to have a broad range of unintended effects. Some of these effects are caused by the method of gene transfer that escapes the plants' own gene regulation. There is a broad range of relevant issues such as cumulative effects and synergisms, genome-environment interactivity as well

contaminations with viable material. Many examples of unintended effects of genetically engineered plants can hardly be detected at an early stage of risk assessment (Then 2010a, 4, 9).

Chapter 7

1   One can discern a striking dual narrative about technology. The technology is portrayed as brand new and capable of creating a new society through genetic modification when technological elites speak to investors, policymakers or patent offices, and to publics ready to be enrolled in the new venture. But when actual or anticipated concerns need to be assuaged, the same technology is presented as nothing unusual (i.e., we have been modifying genetic makeup of organisms all the time) (Felt and Wynne 2007, 26).

2   These claims have not been accepted by GMO skeptics/opponents. There are studies that directly challenge the purported qualities of GMOs (Antoniou, Robinson and Fagan 2012; Glover 2009, 2010; Paul et al. 2003).

3   The irony here is that by promoting GMOs as a technical fix for the socio-environmental damage done by earlier technologies, agribiotech companies in essence admit that technological interventions can have unintended and undesired side effects. This concession comes, of course, in stark contrast with their claim that GMOs are a perfectly controllable and predictable technology.

4   While this notion was discussed in Chapter 2, we can now elaborate on it a bit further through the prism of nature and society relations.

5   In the Rio Declaration, Principle 15 states:

In order to protect the environment, the precautionary approach shall be widely applied by States according to their capabilities. Where there are threats of serious or irreversible damage, lack of full scientific certainty shall not be used as a reason for postponing cost-effective measures to prevent environmental degradation. (UNESCO 1992, 3)

The subsequent Cartagena Protocol reaffirmed "the precautionary approach contained in Principle 15 of the Rio Declaration on Environment and Development" (Secretariat of the Convention on Biological Diversity 2000, 2–3). Building on the Cartagena Protocol, the European Commission has set out two specific cases where the principle is applicable:

- "Where the scientific data are insufficient, inconclusive or uncertain;
- "Where a preliminary scientific evaluation shows that potentially dangerous effects for the environment and human, animal or plant health can reasonably be feared.

"In both cases, the risks are incompatible with the high level of protection sought by the European Union. [The Commission has also set out] three rules which need to be followed for the precautionary principle to be observed:

- A complete scientific evaluation carried out by an independent authority in order to determine the degree of scientific uncertainty;
- An assessment of the potential risks and the consequences of inaction;
- The participation, under conditions of maximum transparency, of all the interested parties in the study of possible measures.

"[T]he precautionary principle may take the form of a decision to act or not to act, depending on the level of risk considered 'acceptable'. The Union had applied this precautionary principle in the area of genetically modified organisms (GMOs), for instance, with the adoption of a moratorium on their commercialisation between 1999 and 2004" (Europa 2013).

In addition, in the 2001/18 EU Directive it is clearly stated:

> The precautionary principle has been taken into account in the drafting of this Directive and must be taken into account when implementing it [...] In accordance with the precautionary principle, the objective of this Directive is to approximate the laws, regulations and administrative provisions of the Member States and to protect human health and the environment when:
>
> - Carrying out the deliberate release into the environment of genetically modified organisms for any other purposes than placing on the market within the Community,
> - Placing on the market genetically modified organisms as or in products within the Community (European Commission 2001, 1–4).

6   In the EU, the burden of proof is placed on the shoulders of the "proto-developer," who has to demonstrate that the new technology poses no (reasonable) environmental harm (O'Riordan and Jordan 1995, 5).

7   While we argued in Chapter 5 that R&D and Farming constitute distinct social locations, in our case where the GMO is cultivated only as part of the testing phase (i.e., there are no farmers cultivating the plant for commercial purposes), we examined these two cases under one category.

Chapter 8

1   The Canadian regulatory system is a case in point. Under this "trait-based" system, GM crops undergo full risk assessment only if they are defined as being a "plant with a novel trait"—that is, only if they display a characteristic not seen in another product already approved. ACRE, BBSRC, CST, EASAC and STC have all advocated a move to such a trait-based regulatory system as a response to the current stalemate in the EU (House of Commons Science and Technology Committee 2015a, 35–36). The viewpoint that such a trait-based regulatory system will make the approval process more efficient was, nonetheless, challenged by Elisabeth Waigmann (head of the GMO unit in EFSA) and Joe Perry (chair of the GMO Panel of EFSA) in their oral evidence to the STC Inquiry (House of Commons Science and Technology Committee 2014b, Q330).

2   For a balanced account of the "Séralini affair" see Krimsky (2015, 901–6).

3   In brief, this was the first life-long rodent feeding study investigating possible toxic effects rising from Monsanto Roundup herbicide and Roundup-tolerant GM maize. In this highly controversial study, Séralini and colleagues found alarming effects on rats including "liver congestions and necrosis [...] 2.5–5.5 times higher [than in the control group] (4221), "large tumor incidence by 2–3-fold in comparison to [the] controls" (4228, 29), "large mammary tumors" (4229) and "hormone-dependent mammary, hepatic and kidney disturbances" (4230).

4   The most widely known collection of herbicide-resistant crops is Monsanto's "Roundup ready" products, which exhibit tolerance to Roundup, the company's widely

commercialized glyphosate-based weed-killer. As discussed in Chapter 6, the former director of Rothamsted Research, Maurice Moloney, was responsible for the development of Roundup Ready Canola.

5   "Golden Rice" is a variety of rice genetically engineered to produce beta-carotene (pro-vitamin A) to help combat vitamin A deficiency (Paine et al. 2005). "Vitamin A deficiency (VAD) is the leading cause of preventable blindness in children and increases the risk of disease and death from severe infections. In pregnant women VAD causes night blindness and may increase the risk of maternal mortality. Vitamin A deficiency is a public health problem in more than half of all countries, especially in Africa and South-East Asia, hitting hardest young children and pregnant women in low-income countries" (WHO 2016).

6   Australia, Canada and the United States were the three countries that did not fully approve the *Global Report*.

7   It is worth noting that in August 1999, the UK's Advertising Standards Authority (ASA) concluded that Monsanto's advertisements about GM foods and crops were "misleading" in claiming, among other erroneous allegations regarding safety tests and approvals of GM foods, that genetic modification was an extension of traditional breeding methods (Hall 1999).

8   This problem is similar to what Collins has termed "the experimenter's regress." The experimenter's regress in based on the observation that "experiments, especially those on the frontiers of science, are difficult and that there is no criterion, other than the outcome, that indicates whether the difficulties have been overcome" (Collins 2005, 457). In other words: "We won't know if we have built a good detector until we have tried it and obtained the correct outcome. But we don't know what the correct outcome is until [...] and so on *ad infinitum*" (Collins and Pinch 1998, 98).

9   Since Sjöstedt leaves open the usage of the words *conflict* and *dispute* (2009, 227), I find the latter more appropriate in the case of GMOs, as it implies a milder and more limited confrontational orientation.

BIBLIOGRAPHY

Abercrombie, Nicholas and Bryan S. Turner. 1978. "The Dominant Ideology Thesis." *The British Journal of Sociology* 29, no. 2: 149–70.

Agricultural Biotechnology Council. 2012a. "Going for Growth." *ABC*. Accessed October 16, 2012. http://tinyurl.com/goingforgrowth.

———. 2012b. "Going for Growth Roundtable Feedback." *Genewatch*. Accessed December 4, 2012. http://www.genewatch.org/uploads/f03c6d66a9b354535738483c1c3d49e4/Going_for_Growth_roundtable___summary.pdf.

Alexander, Jeffrey. 1998. *Neofunctionalism and After*. Oxford: Blackwell.

Althusser, Louis. 1969. *For Marx*. London: Penguin.

———. 1971. *Lenin and Philosophy and Other Essays*. London: New Left Books,.

Altieri, Miguel A. 2005. "The Myth of Coexistence: Why Transgenic Crops Are Not Compatible with Agroecologically Based Systems of Production." *Bulletin of Science Technology & Society* 25, no. 4: 361–71.

Amis, Anthony. 2010. "Submission from Friends of the Earth into the National Registration Scheme for Agricultural and Veterinary Chemicals." Australian Government: Department of Agriculture, Fisheries and Forestry. Accessed May 23, 2013. http://www.daff.gov.au/agriculture-food/ag-vet-chemicals/domestic-policy/naspt/responses-to-discussion-paper/friends_of_the_earth.

Ammann, Klaus and Marcel Kuntz. 2016. Decades-Old GMO Regulation Unfit for 21st Century. January 12. Accessed January 13, 2016. http://www.euractiv.com/sections/agriculture-food/decades-old-gmo-regulation-unfit-21st-century-320888.

Amos, Ilona. 2015. "Scots GM Crops Backing 'Like a Religious Crusade.'" August 19. Accessed January 24, 2016. http://www.scotsman.com/news/environment/scots-gm-crops-backing-like-a-religious-crusade-1-3862310.

Antoniou, Michael et al. 2011. "Roundup and Birth Defects: Is the Public Being Kept in the Dark?" *Earth Open Source*. Accessed October 10, 2015 www.earthopensource.org/files/pdfs/Roundup-and-birth-defects/RoundupandBirthDefectsv5.pdf.

Antoniou, Michael, Claire Robinson and John Fagan. 2012. "GMO Myths and Truths." Vers. 1.3b. Earth Open Source. Accessed April 2, 2013. www.earthopensource.org.

AquaBounty Technologies. 2013. AquAdvantage® Fish. Accessed June 11, 2013. https://aquabounty.com/innovation/technology/.

Archer, Margaret. 1996. *Culture and Agency: The Place of Culture in Society*. Rev. ed. Cambridge: Cambridge University Press.

Avé, Dirk A., Peter Gregory and Ward Tingey. 1987. "Aphid Repellent Sesquiterpenes in Glandular Trichomes of Solanum Berthaultii and S. Tuberosum." *Entomologia Experimentalis et Applicata* 44: 131–38.

Axelrad, Jeffrey C., Vyvyan C. Howard and Graham W. McLean. 2003. "The Effects of Acute Pesticide Exposure on Neuroblastoma Cells Chronically Exposed to Diazinon." *Toxicology* 185: 67–78.

BASF. 2014. *Written Evidence Submitted by BASF plc GMC0019*. May 7. Accessed October 30, 2015. http://data.parliament.uk/writtenevidence/committeeevidence.svc/evidencedocument/science-and-technology-committee/gm-foods-and-application-of-the-precautionary-principle-in-europe/written/8586.pdf.

Bauer-Panskus, Andreas and Christoph Then. 2014. "Comments Regarding the GRACE Publication 'Ninety-Day Oral Toxicity Studies on Two Genetically Modified Maize MON810 Varieties in Wistar Han RCC Rats EU 7th Framework Programme Project GRACE.'" *Testbiotech*. November 7. Accessed January 24, 2016. http://www.grace-fp7.eu/sites/default/files/Testbiotech_Doubts_%20EU_Reseach_Project_GRACE_2.pdf.

Baulcombe, David, Jim Dunwell, Jonathan Jones, John Pickett and Pere Puigdomenech. 2014. *GM Science Update: A Report to the Council for Science and Technology*. CST/14/634a. *UK Government*. March 14. Accessed November 22, 2015. https://www.gov.uk/government/uploads/system/uploads/attachment_data/file/292174/cst-14-634a-gm-science-update.pdf.

BBC. 1999. "No Regrets Over GM Protest." Accessed April 30, 2013. http://news.bbc.co.uk/2/hi/uk_news/406191.stm.

———. 2003. "BBC News in Depth: Microsoft." Accessed January 25, 2011. http://news.bbc.co.uk/1/hi/in_depth/business/2000/microsoft/default.stm.

———. 2004. "Monsanto Drops Plans for GM Wheat." Accessed December 4, 2012. http://news.bbc.co.uk/2/hi/business/3702739.stm.

———. 2009. "Heart Surgery "More Successful." Accessed September 5, 2013 http://news.bbc.co.uk/2/hi/health/8170618.stm.

BBSRC. 2009. "The Age of Bioscience: Strategic Plan 2010–2015." Accessed April 1, 2013. http://www.bbsrc.ac.uk/web/FILES/Publications/strategic_plan_2010-2015.pdf.

———. 2010. "New Director for Rothamsted Research." Accessed September 11, 2012. http://www.bbsrc.ac.uk/news/people-skills-training/2010/100114-pr-new-director-for-rothamsted-research.aspx.

———. 2012. *Our Organisation*. Accessed September 9, 2012. http://www.bbsrc.ac.uk/organisation/mission.aspx.

———. 2012a. *Council Conflicts*. Accessed September 10, 2012. http://www.bbsrc.ac.uk/web/FILES/Conflicts/council_conflicts.pdf.

———. 2012b. *News and Events*. Accessed September 9, 2012. http://www.bbsrc.ac.uk/news/food-security/2012/120330-pr-rothamsted-aphid-resistant-wheat-trial.aspx.

———. 2015. "GM, Synthetic Biology and Genome Editing." December 3. Accessed January 19, 2016. http://www.bbsrc.ac.uk/research/briefings/gm-synthetic-biology-genome-editing.

Beck, Ulrich, Anthony Giddens and Scott Lash. 1994. *Reflexive Modernization: Politics, Tradition and Aesthetics in the Modern Social Order*. Cambridge: Polity.

Benbrook, Charles M. 2012. "Impacts of Genetically Engineered Crops on Pesticide Use in the U.S.—The First Sixteen Years." *Environmental Sciences Europe* 24, no. 24: 1–13.

Beniger, James. 1986. *The Control Revolution: Technological and Economic Origins of the Information Society*. Cambridge: Harvard University Press.

Bennett, Heather. 2009. "Canola Trials Come a Cropper." January 16. Accessed January 12, 2016. https://www.businessnews.com.au/article/GM-canola-trials-come-a-cropper.

Benton, Ted, ed. 1977. *Philosophical Foundations of the Three Sociologies*. London: Routledge & Kegan Paul.

————. 1984. *The Rise and Fall of Structural Marxism*. London and Basingstoke: Macmillan.

————. 1991. "Biology and Social Science: Why the Return of the Repressed Should Be Given a Cautious Welcome." *Sociology* 25, no. 1: 1–29.

————. 1993. *Natural Relations: Ecology, Animal Rights and Social Justice*. London: Verso.

————. 1994. "Biology and Social Theory in the Environment Debate." In *Social Theory and the Global Environment*, edited by Michael Redclift and Ted Benton, 28–50. London: Routledge.

————, ed. 1996. *The Greening of Marxism*. New York: Guilford.

————. 2006. *Bumblebees New Naturalist no. 98*. London: HarperCollins UK.

————. 2007. "Deep Ecology." In *The Sage Handbook of Environment and Society*, edited by Jules Pretty, Andrew S. Ball, Ted Benton, Julia S. Guivant, David R. Lee, David Orr, Max J. Pfeffer and Hugh Ward, 78–90. London: Sage Publications.

————. 2012. *Grasshoppers and Crickets. Collins New Naturalist Library, Book 120*. London: HarperCollins Publishers.

Benton, Ted and Ian Craib. 2001. *Philosophy of Social Science: The Philosophical Foundations of Social Thought*. Basingstoke: Palgrave.

Bercovitch, Jacob. 2009. "Mediation and Conflict Resolution." In *The Sage Handbook of Conflict Resolution*, edited by Jacob Bercovitch, Victor Kremenyuk and William Zartman, 340–57. London: Sage.

Berger, Peter and Thomas Luckmann. 1967. *The Social Construction of Reality: A Treatise in the Sociology of Knowledge*. London: Penguin Classics.

Bijker, Wiebe and John Law. 1992. *Shaping Technology / Building Society*. Cambridge, MA: The MIT Press.

Bimber, Bruce. 1990. "Karl Marx and the Three Faces of Technological Determinism." *Social Studies of Science* 20, no. 2: 333–51.

BIS. 2012. *UK Government*. October 11. Accessed December 2, 2015. https://www.gov.uk/government/consultations/shaping-a-uk-agri-tech-strategy-call-for-evidence.

————. 2013. *A UK Strategy for Agricultural Technologies*. BIS/13/1060. London: Crown Copyright.

Blair, Tony. 2002. "Science Matters." *Life* 54: 155–59.

Bloor, David. 1976. *Knowledge and Social Imagery*. London: Routledge & Kegan Paul.

Bottomore, Tom, ed. 1991. *A Dictionary of Marxist Thought*. 2nd ed. London: Blackwell Publishing.

Bourdieu, Pierre. 1977. *Outline of a Theory of Practice*. Cambridge: Cambridge University Press.

————. 1990. *The Logic of Practice*. Cambridge: Polity Press.

Bové, José. 2001. "A Movement of Movements?" *New Left Review* no. 12: 89–101.

Breslau, Daniel. 2000. "Sociology after Humanism: A Lesson from Contemporary Science Studies." *Sociological Theory* 18, no. 2: 289–307.

Brian, John. 2012. "Revealed: Rothamsted Scientist's Role in Destruction of Key GM Research." May 20. Accessed December 18, 2015. http://www.gmfreecymru.org/news/Press_Notice20May2012.htm.

Britt, Bailey and Marc Lappé. 2002. *Engineering the Farm: Ethical and Social Aspects of Agricultural Biotechnology*. Washington, DC: Island Press.

Brookers, Graham and Peter Barfoot. 2011. "GM Crops: Global Socio-economic and Environmental Impacts 1996–2009." PG Economics Ltd, Dorchester.

BSPP. 2012. "Professor Graham Jellis." Accessed December 1, 2012. http://www.bspp.org.uk/profiles/jellis.php.

Callon, Michel. 1986. "Some Elements of a Sociology of Translation: Domestication of the Scallops and the Fishermen of St. Brieuc Bay." In *Power, Action and Belief: A New Sociology of Knowledge?*, edited by John Law, 196–223. London: Routledge & Kegan Paul.

———. 1999. "Actor-Network Theory—The Market Test." In *Actor Network Theory and After*, edited by John Law and John Hassard, 181–95. Oxford: Blackwell Publishing.

Carrell, Severin. 2015. "Scotland to Issue Formal Ban on Genetically Modified Crops." August 9. Accessed November 7, 2015. http://www.theguardian.com/environment/2015/aug/09/scotland-to-issue-formal-ban-on-genetically-modified-crops.

Case, Philip. 2012a. *DEFRA Minister Backs Use of "Safe" GM Crops.* November 22. Accessed November 7, 2015. http://www.fwi.co.uk/arable/defra-minister-backs-use-of-safe-gm-crops.htm.

———. 2012b. *GM Wheat Trial Begins Amid High Security.* Accessed November 30, 2012. http://www.fwi.co.uk/arable/gm-wheat-trial-begins-amid-high-security.htm.

Catacora-Vargas, Georgina. 2014. *Sustainability Assessment of Genetic ally Modified Herbicide Tolerant Crops.* Biosafety Report 2014/02, Tromsø: GenØk Centre for Biosafety.

Chan, Christine, Toity Deave and Trisha Greenhalgh. 2010. "Childhood Obesity in Transition Zones: An Analysis Using Structuration Theory." *Sociology of Health and Illness* 32, no. 5: 711–29.

Channel 4. *Protest Over GM Wheat Study.* 2012. Accessed December 1, 2012. http://www.channel4.com/news/protest-over-gm-wheat-study.

Charles, Daniel. *Lords of the Harvest: Biotech, Big Money, and the Future of Food.* Cambridge: Basic Books, 2001.

Clapp, Jennifer. 2007. "Transnational Corporate Interests in International Biosafety Negotiations." In *The International Politics of Genetically Modified Food*, by Robert Falkner, 34–47. New York: Palgrave Macmillan.

Clover, Charles. 2012. *The Tendrils of GM Inch in by the Back Door.* Accessed September 2, 2012. http://www.thesundaytimes.co.uk/sto/comment/columns/charlesclover/article1031782.ece.

Clutterbuck, Charles. 2014. "Charles Clutterbuck—written evidence." October 14. House of Commons Select Committee. Accessed November 5, 2015. <http://data.parliament.uk/writtenevidence/committeeevidence.svc/evidencedocument/science-and-technology-committee/gm-foods-and-application-of-the-precautionary-principle-in-europe/written/13740.pdf>.

Coad, Alan and Ian Herbert. 2009. "Back to the Future: New Potential for Structuration Theory in Management Accounting Research?" *Management Accounting Research* 20: 177–92.

Cobb, Roger W. and Charles D. Elder. 1971. "The Politics of Agenda-Building: An Alternative Perspective for Modern Democratic Theory." *The Journal of Politics* 33, no. 4: 792–915.

Cohen, Gerald A. 1978. *Karl Marx's Theory of History: A Defence.* London: Oxford University Press.

Cohen, Ira J. 1996. "Reviewed work(s): Sociological Theory: What Went Wrong? Diagnosis and Remedies. by Nicos Mouzelis." *The American Journal of Sociology* 102, no. 2: 595–96.

Collins, Harry. 1981a. "Introduction: Stages in the Empirical Programme of Relativism." *Social Studies of Science* 11, no. 1, Special Issue: *Knowledge and Controversy: Studies of Modern Natural Science.* Feb: 3–10.

———. 1981b. "'Son of Seven Sexes': The Social Destruction of a Physical Phenomenon." *Social Studies of Science* 11: 33–62.

———. 1983. "The Sociology of Scientific Knowledge: Studies of Contemporary Science." *Annual Review of Sociology* 9: 265–85.

———. 1992. *Changing Order: Replication and Induction in Scientific Practice.* Chicago: University of Chicago Press.

———. 2005. "Replication." In *Science, Technology, and Society: An Encyclopedia,* edited by Sal Restivo, 456–58. Oxford: Oxford University Press.

Collins, Harry, and Graham Cox. 1976. "Recovering Relativity: Did Prophecy Fail?" *Social Studies of Science* 6, no. 3/4: 423–44.

Collins, Harry, and Steve Yearley. 1992. "Epistemological Chicken." In *Science as Practice and Culture,* edited by Andrew Pickering, 301–25. London: University of Chicago Press.

Collins, Harry, and Trevor Pinch. 1998. *The Golem: What You Should Know about Science.* 2nd ed. Cambridge: Cambridge University Press.

———. 2005. *Dr. Golem: How to Think about Medicine.* Chicago: The University of Chicago Press.

Collins, Randall. 1981. "On the Microfoundations of Macrosociology." *American Journal of Sociology* 86, no. 5: 984–1014.

Committee, Commons Select. 2014 *GM Foods and Application of the Precautionary Principle in Europe: Terms of Reference.* February 14. Accessed November 9, 2015. http://www.parliament. uk/business/committees/committees-a-z/commons-select/science-and-technology-committee/news/140214-gm-foods-and-application-of-the-precautionary-principle-in-europe.

Conko, Gregory, and Henry I. Miller. 2001. "Precaution without Principle." *Nature Biotechnology* 19, no. 4: 302–3.

Connor, Steve. 1999. *Scientists Revolt at Publication of "Flawed" GM Study.* October 11. Accessed December 18, 2015. http://www.independent.co.uk/environment/scientists-revolt-at-publication-of-flawed-gm-study-737888.html.

———. 2012. *The Independent: Science.* Accessed September 2, 2012. http://www.independent.co.uk/news/science/steve-connor-opponents-of-this-crop-trial-are-blind-to-the-food-crisis-7792517.html.

Corporate Europe Observatory and Earth Open Source. 2012. "Conflicts on the Menu." *Earth Open Source.* Accessed November 27, 2012. http://earthopensource. org/files/pdfs/Conflicts_on_the_menu_report/Conflicts_on_the_menu_report_English.pdf.

Council for Science and Technology. 2014a. "Council for Science and Technology." *CST Reports on Science and Technology.* March 14. Accessed November 30, 2015. https://www. google.gr/url?sa=t&rct=j&q=&esrc=s&source=web&cd=1&cad=rja&uact=8&ved= 0ahUKEwiTkf7whLnJAhWInRoKHZZPBYwQFggeMAA&url=https%3A%2F%2 Fwww.gov.uk%2Fgovernment%2Fuploads%2Fsystem%2Fuploads%2Fattachment_ data%2Ffile%2F288823%2Fcst-14-634-gm-technologie.

———. 2014b. "GM Technologies." *UK Government.* March 14. Accessed November 30, 2015. https://www.gov.uk/government/uploads/system/uploads/attachment_data/ file/288823/cst-14-634-gm-technologies.pdf.

Craib, Ian. 1981. "'Criticism and Ideology:' Theory and Experience." *Contemporary Literature* 22, no. 4, *Marxism and the Crisis of the World*: 489–509.

CropLife International. 2013. *Climate Change.* Accessed April 10, 2013. http://actionforag. org/issue/climate-change.

———. 2016. *A Seed Story: Plant Biotechnology in Focus.* February 7. Accessed February 7, 2016. http://croplife.org/biotech-crop-development.

Culp, Sylvia. 1995. "Objectivity in Experimental Inquiry: Breaking Data-Technique Circles." *Philosophy of Science* 62: 430–50.

Dafoe, Allan. 2015. "On Technological Determinism: A Typology, Scope Conditions, and a Mechanism." *Science, Technology, and Human Values* 40, no. 6: 1047–76.

de Vendômois, Joël S, Dominique Cellier, Christian Vélot, Emilie Clair, Robin Mesnage and Gilles-Eric Séralini. 2010. "Debate on GMOs Health Risks after Statistical Findings in Regulatory Tests." *International Journal of Biological Sciences* 6, no. 6: 590–98.

De Vos, Martin, Wing Yin Cheng, Holly E. Summers, Robert A. Raguso, and Georg Jander. 2010. "Alarm Pheromone Habituation in Myzus Persicae has Fitness Consequences and Causes Extensive Gene Expression Changes." *PNAS* 107, no. 33: 14673–78.

Decker, Susan. 2013. "Apple Import Ban on Old IPhones Stokes Samsung Patent War." Accessed July 16, 2013. http://www.bloomberg.com/news/2013-06-04/apple-faces-u-s-import-ban-on-some-devices-after-samsung-win.html.

DEFRA. 2011. "The Environment: DEFRA." *DEFRA*. Accessed July 5, 2012. http://archive.defra.gov.uk/environment/quality/gm/regulation/registers/consents/index.htm.

Department of Trade and Industry. 2003. "GM Nation? The Findings of the Public Debate." London.

Diamand, Emily. 2001. "Great Food Gamble." Friends of the Earth. Accessed October 13, 2012. http://www.foe.co.uk/resource/reports/great_food_gamble.pdf.

Diels, Johan, Mario Cunha, Célia Manaia, Sabugosa-Madeira Bernardo and Margarida Silva. 2011. "Association of Financial or Professional Conflict of Interest to Research Outcomes on Health Risks or Nutritional Assessment Studies of Genetically Modified Products." *Food Policy* 36: 197–203.

Directorate-General for Research and Innovation. 2010. *A Decade of EU-Funded GMO Research 2001–2010*. EUR24473, Luxembourg: European Commission.

Doing, Park. 2008. "Give Me a Laboratory and I Will Raise a Discipline: The Past, Present, and Future Politics of Laboratory Studies in STS." In *The Handbook of Science and Technology Studies*, edited by Edward Hackett, Olga Amsterdamska, Michael Lynch and Judy Wajcman, 279–96. Hong Kong: MIT Press.

Dreyer, Marion, Ortwin Renn, Adrian Ely, Andy Stirling, Ellen Vos, and Frank Wendler. 2008. *A General Framework for the Precautionary and Inclusive Governance of Food Safety in Europe*. Final Report of subproject 5 of the EU Integrated Project SAFE FOODS, Stuttgart: DIALOGIK.

Driver, Alistair. 2012. "Kendall Likens GM Activists to 1930s Nazis." Accessed December 2, 2012. http://www.farmersguardian.com/home/arable/rothamsted-vandal-causes-significant-damage-at-gm-site/47112.article.

DW. 2015. *Nineteen EU Countries Seek GMO Opt-Out*. October 4. Accessed November 2, 2015. http://www.dw.com/en/nineteen-eu-countries-seek-gmo-opt-out/a-18760718.

Eagleton, Terry. 1979. "Ideology, Fiction, Narrative." *Social Text* 2: 62–80.

———. 1991. *Ideology*. London: Verso.

Earls, Elly. 2011. *Genetically Modifying the EU Food Chain*. Accessed February 6, 2013. http://www.foodprocessing-technology.com/features/featuregenetically-modifying-the-eu-food-chain.

Eckersley, Robyn. 1992. *Environmentalism and Political Theory: Toward an Ecocentric Approach*. Albany, NY: State University of New York Press.

EcoNexus. 2010. *Who We Are*. Accessed March 29, 2013. http://www.econexus.info/who-we-are.

————. 2013. *What is EcoNexus?* Accessed March 28, 2013. http://www.econexus.info/what-econexus.

EFSA. 2004. "Guidance Document of the Scientific Panel on Genetically Modified Organisms for the Risk Assessment of Genetically Modified Plants and Derived Food and Feed." *EFSA Journal* 99: 1–94.

————. 2010. "Guidance on the Environmental Risk Assessment of Genetically Modified Plants." *EFSA Journal* 8, no. 11: 1879–1990.

————. 2011. "Guidance for Risk Assessment of Food and Feed from Genetically Modified Plants." *EFSA Journal* 9, no. 5: 2150–87.

————. 2012a. "FAQ on the Resignation of Diana Banati as Member and Chair of EFSA's Management Board." Accessed February 10, 2013. http://www.efsa.europa.eu/en/faqs/faqresignationdianabanati.htm.

————. 2012b. "Modification of the Existing MRL for Glyphosate in Lentils." *EFSA Journal* 10, no. 1: 2550–75.

————. 2012c. "Decision of the Executive Director implementing EFSA's Policy on Independence and Scientific Decision-Making Processes regarding Declarations of Interests" [Pdf Document]. Accessed January 29, 2013, http://www.efsa.europa.eu/en/keydocs/docs/expertselection.pdf.

————. 2012d. FAQ on Genetically Modified Organisms. Accessed January 29, 2013. http://www.efsa.europa.eu/en/faqs/faqgmo.htm

————. 2012f. Timeline. Accessed January 21, 2013. http://www.efsa.europa.eu/en/10thanniversary/timeline.htm

————. 2012g. What We Do. Accessed January 22, 2012. http://www.efsa.europa.eu/en/aboutefsa/efsawhat.htm

————. 2013. *Organisational Structure.* Accessed January 24, 2013. http://www.efsa.europa.eu/en/efsawho/efsastructure.htm.

————. 2014. "Risk Assessment vs. Risk Management: What's the Difference?" April 16. Accessed December 29, 2015. http://www.efsa.europa.eu/en/press/news/140416.

————. 2015a. *Genetically Modified Organisms.* July 1. Accessed January 30, 2016. http://www.efsa.europa.eu/en/topics/topic/gmo.

————. 2015b. *Organisational Structure.* November 3. Accessed December 20, 2015. http://www.efsa.europa.eu/en/people/orgstructure.

EFSA Panel on Genetically Modified Organisms. 2015a. "Guidance for Renewal Applications of Genetically Modified Food and Feed Authorised under Regulation EC No 1829/2003." *EFSA Journal* 13, no. 6: 4129–37.

————. 2015b. "Guidance on the Agronomic and Phenotypic Characterisation of Genetically Modified Plants." *EFSA Journal* 13, no. 6: 4128–72.

Eisenberg, Rebecca S. 1987. "Proprietary Rights and the Norms of Science in Biotechnology Research." *The Yale Law Journal* 97, no. 2: 177–231.

Ellis, Jaye. 2006. "Overexploitation of a Valuable Resource? New Literature on the Precautionary Principle." *The European Journal of International Law* 17, no. 2: 445–62.

Ellstrand, Norman C. 2001. "When Transgenes Wander, Should We Worry?" *Plant Physiology* 125: 1543–45.

Ellul, Jacques. 1980. *The Technological System.* New York: Continuum.

Elmore, Roger et al. 2001. "Glyphosate-Resistant Soybean Cultivar Yields Compared with Sister Lines." *Agronomy Journal* 93: 408–12.

Eriksson, Mikael, Lennart Hardell, Michael Carlberg, and Måns Akerman. 2008. "Pesticide Exposure as Risk Factor for Non-Hodgkin Lymphoma Including Histopathological Subgroup Analysis." *International Journal of Cancer* 123: 1657–63.

Euractiv. 2006. *WTO panel rules EU GMO moratorium illegal.* Accessed December 5, 2012. http://www.euractiv.com/trade/wto-panel-rules-eu-gmo-moratorium-illegal/article-152341.

———. 2009. "Germany Joins Ranks of Anti-GMO Countries." April 15. Accessed February 5, 2016. http://www.euractiv.com/cap/germany-joins-ranks-anti-gmo-cou-news-221725.

———. 2012a. "Dalli Resignation Row Turns to War of Words with Barroso." Accessed February 10, 2013. http://www.euractiv.com/health/dalli-barroso-war-words-news-515653.

———. 2012b. "EU Agencies Stained by 'Conflicts of Interest', Wrongdoing." Accessed February 10, 2013. http://www.euractiv.com/euro-finance/eu-agencies-marred-conflict-inte-news-512587.

———. 2012c. "EU Science Advisor: 'Lots of Policies Are Not Based on Evidence.'" *Euractiv.* Accessed October 13, 2012. http://www.euractiv.com/innovation-enterprise/chief-scientifc-adviser-policy-p-interview-514074.

———. 2012d. "No Risk with GMO Food, Says EU Chief Scientific Advisor." Accessed February 6, 2013. http://www.euractiv.com/innovation-enterprise/commission-science-supremo-endor-news-514072.

———. 2012e. "Without Dalli, GMO Foes Hope for Tougher EU Policy." Accessed February 10, 2013. http://www.euractiv.com/science-policymaking/dalli-groups-bank-tougher-gmo-st-news-515963.

———. 2013a. *GMOs:* "We Shouldn't Mix the Precautionary Principle and Public Perception." Accessed May 31, 2013. http://www.euractiv.com/cap/gmos-shouldn-mix-precautionary-principle-public-perception/article-168021.

———. 2013b. "UK Minister Urges EU to Speed Up GM Crop Approvals." Accessed January 31, 2013. http://www.euractiv.com/development-policy/uk-minister-urges-eu-speed-gm-cr-news-516852?utm_source=EurActiv%20Newsletter&utm_campaign=4aa1c066f8-newsletter_agriculture__food&utm_medium=email.

Europa. 2013. *Precautionary Principle.* Accessed June 6, 2013. http://europa.eu/legislation_summaries/glossary/precautionary_principle_en.htm.

EuropaBio. 1999. "Discussion Paper on the Precautionary Principle." EuropaBio. Accessed May 31, 2013. http://www.europabio.org/sites/default/files/position/discussion_paper_on_the_precautionary_principle.pdf.

———. 2009. "Position Papers and Publications: Agricultural Biotech." EuropaBio: The European Association of Bioindustries. November 20. Accessed June 1, 2010. http://www.europabio.org/positions/GBE/PP_101209_basicsbiotech.pdf.

———. 2010a. "Position Papers and Publications: Agricultural Biotech." The European Association of Bioindustries. EuropaBio Website. January 12. Accessed June 1, 2010. http://www.europabio.org/positions/GBE/PP_080110-Socio-economic-impacts-of-GM-Crops-GMO.pdf.

———. 2010b. "Socio-economic Impacts of Green Biotechnology." EuropaBio. Accessed April 10, 2013. http://www.europabio.org/sites/default/files/report/pp_socio-economic-impacts-of-gmo.pdf.

———. 2011a. *Approvals of GMOs in the European Union.* Accessed January 31, 2013. http://www.europabio.org/sites/default/files/report/approvals_of_gmos_in_eu_europabio_report.pdf.

———. 2011b. *Approvals of GMOs in the European Union.* Accessed February 4, 2013. http://www.europabio.org/approvals-gmos-european-union.

————. 2012a. *Anne Glover: "It is High Time to Be Less Modest."* Accessed December 6, 2012. http://www.europabio.org/agricultural/news/anne-glover-it-high-time-be-less-modest.

————. 2012b. "DEFRA Minister Backs Use of 'Safe' GM Crops." Accessed December 5, 2012. http://www.europabio.org/agricultural/news/defra-minister-backs-use-safe-gm-crops.

————. 2012c. *News.* Accessed November 28, 2012. http://www.europabio.org/agricultural/news/defra-minister-backs-use-safe-gm-crops.

————. 2012d. "Pocket Guide to GM Crops and Policies." EuropaBio. Accessed November 11, 2012. http://www.europabio.org/sites/default/files/position/pocket_guide_gmcrops_policy.pdf.

————. 2013a. *How We Are Organised.* Accessed September 4, 2013. http://www.europabio.org/how-we-are-organised.

————. 2013b. *Members.* Accessed September 4, 2013. http://www.europabio.org/members.

————. 2013c. *Science Not Fiction: Time to Think Again about GM.* EuropaBio. January 23. Accessed April 10, 2013. http://seedfeedfood.eu/wp-content/uploads/2013/02/flip-book2.pdf.

European Commission. 2001. "Directive 2001/18/EC of the European Parliament and of the Council of 12 March 2001 on the Deliberate Release into the Environment of Genetically Modified Organisms and Repealing Council Directive 90/220/EEC." *Official Journal of the European Communities* L, no. 106: 1–38.

————. 2002. "Communication from the Commission to the Council, the European Parliament, the Economic and Social Committee and the Committee of the—Life Sciences and Biotechnology—A Strategy for Europe." *EUR-Lex: Access to European Union Law.* Accessed June 20, 2013. http://eur-lex.europa.eu/LexUriServ/LexUriServ.do?uri=COM:2002:0027:FIN:EN:PDF.

————. 2003a. "European Commission Regrets US Decision to file WTO Case on GMOs as Misguided and Unnecessary." European Commission. Accessed December 6, 2012. http://trade.ec.europa.eu/doclib/docs/2003/november/tradoc_114679.pdf.

————. 2003b. "Guidelines for the Development of National Strategies and Best Practices to Ensure the Co-existence of Genetically Modified Crops with Conventional and Organic Farming." European Commission. Accessed June 14, 2013. http://ec.europa.eu/agriculture/publi/reports/coexistence2/index_en.htm.

————. 2003c. "Regulation (EC) No 1830/2003 of the European Parliament and of the Council of 22 September 2003 concerning the traceability and labelling of genetically modified organisms." *Official Journal of the European Union* L(268): 24–28.

————. 2006. *Europeans and Biotechnology in 2005: Patterns and Trends.* Eurobarometer 64.3, Brussels: European Commission,

————. 2009. *Report from the Commission to the Council and the European Parliament on the Coexistence of Genetically Modified Crops with Conventional and Organic Farming.* COM 2009 153 final, Brussels: European Commission.

————. 2010a. *Eurobarometer 73.1: Biotechnology.* Eurobarometer, Brussels: TNS Opinion and Social,

————. 2010b. "Public Opinion." European Commission. TNS Opinion and Social. Accessed December 4, 2012. http://ec.europa.eu/public_opinion/archives/ebs/ebs_341_cn.pdf.

————. 2011. *Ombudsman: EFSA Should Strengthen Procedures to Avoid Potential Conflicts of Interest in "Revolving Door" Cases.* Accessed February 12, 2013. http://europa.eu/rapid/press-release_EO-11-20_en.htm.

————. 2012a. *Deliberate Release and Placing on the EU Market of GMOs—GMO Register.* April 26. Accessed January 20, 2016. http://gmoinfo.jrc.ec.europa.eu/overview/dbcountries.asp.

————. 2012b. *Existing Rules on GM Food & Animal Feed.* Accessed December 6, 2012. http://ec.europa.eu/food/plant/gmo/legislation/gm_food_animal_feed_en.htm

————. 2012c. *Future Rules—New Approach.* Accessed December 6, 2012. http://ec.europa.eu/food/plant/gmo/legislation/future_rules_en.htm.

————. 2012d. *Legislation on GMO Cultivation.* Accessed December 1, 2012. http://ec.europa.eu/food/plant/gmo/legislation/index_en.htm.

————. 2013. "Commission Implementing Regulation EU No 503/2013 of 3 April 2013." *Official Journal of the European Union* L 157/1: 1–48.

————. 2015a. *Fact Sheet: Questions and Answers on EU's policies on GMOs.* April 22. Accessed December 20, 2015. http://europa.eu/rapid/press-release_MEMO-15-4778_en.htm.

————. 2015b. *New Plant Breeding Techniques.* October 10. Accessed February 16, 2016. http://ec.europa.eu/food/plant/gmo/legislation/plant_breeding/index_en.htm.

————. 2015c. *Review of the Decision-making Process on GMOs in the EU: Questions and Answers.* April 22. Accessed December 20, 2015. http://europa.eu/rapid/press-release_MEMO-15-4779_en.htm.

————. 2015d. *Traceability and Labelling.* October 16. Accessed January 29, 2016. http://ec.europa.eu/food/plant/gmo/traceability_labelling/index_en.htm.

European Commission: Directorate-General for Research and Innovation. 2014. "GRACE: Letter from the EU Commission to Testbiotech December 17, 2014." *Testbiotech.* December 17. Accessed January 24, 2016. http://www.testbiotech.org/sites/default/files/GRACE_Answer_from_EU_Commission_to_Testbiotech.pdf.

European Union. 2005. *Ban on Antibiotics as Growth Promoters in Animal Feed Enters into Effect.* Accessed June 6, 2013. http://europa.eu/rapid/press-release_IP-05-1687_en.htm.

————. 2008. "Consolidated Version of the Treaty on the Functioning of the European Union." *Official Journal of the European Union* 115: 47–199.

————. 2011. Ombudsman: EFSA should strengthen procedures to avoid potential conflicts of interest in 'revolving door' cases. [Online] (Reference: EO/11/20) Accessed February 12, 2013. Available at: http://europa.eu/rapid/press-release_EO-11-20_en.htm.

Fagan, John, Michael Antoniou and Claire Robinson. 2014. *GMO Myths and Truths.* 2nd ed. London: Earth Open Source.

Falkner, Robert. 2000. "Regulating Biotech Trade: The Cartagena Protocol on Biosafety." *International Affairs Royal Institute of International Affairs (1944–)* 76, no. 2, *Special Biodiversity* Issue April: 299–313.

————. 2007. "International Cooperation Against the Hegemon: The Cartagena Protocol on Biosafety." In *The International Politics of Genetically Modified Food,* by Robert Falkner, 15–33. New York: Palgrave Macmillan.

FAO/WHO. 2000. "Safety Aspects of Genetically Modified Foods of Plant Origin: Report of a Joint FAO/WHO Expert Consultation." World Health Organization, Geneva.

FDA. 2013. *Genetically Engineered Salmon.* Accessed June 11, 2013. http://www.fda.gov/AnimalVeterinary/DevelopmentApprovalProcess/GeneticEngineering/GeneticallyEngineeredAnimals/ucm280853.htm.

————. 2015. *FDA Has Determined That the AquAdvantage Salmon is as Safe to Eat as Non-GE Salmon.* November 19. Accessed December 29, 2015. http://www.fda.gov/ForConsumers/ConsumerUpdates/ucm472487.htm.

Felt, Ulrike, and Brian Wynne. 2007. *Taking European Knowledge Society Seriously.* European Commission, Directorate-General for Research Science, Economy and Society.

Fernandez-Cornejo, Jorge, Seth Wechsler, Mike Livingston and Mitchell Lorraine. 2014. *Genetically Engineered Crops in the United States. Economic Research Report Number 162,* Washington, DC: United States Department of Agriculture.

Fine, Gary Alan, and Kent Sandstrom. 1993. "Ideology in Action: A Pragmatic Approach to a Contested Concept." *Sociological Theory* 11, no. 1: 21–38.

Fleming, Jeremy. 2014. *Juncker Still Mulling Scientific Advice Role after Glover's Position Axed.* November 14. Accessed October 21, 2015. http://www.euractiv.com/sections/innovation-enterprise/juncker-still-mulling-scientific-advice-role-309999.

Flynn, Laurie, and Michael Sean Gillard. 1999. *Pro-GM Food Scientist "Threatened Editor."* November 1. Accessed December 18, 2015. http://www.theguardian.com/science/1999/nov/01/gm.food.

Food Standards Agency. 2011. *More about the Steering Group for a Public Dialogue on Food and Use of GM.* August 4. Accessed November 29, 2015. http://tna.europarchive.org/20111023080327/http:/www.food.gov.uk/gmfoods/gm/gmdialogue/moregmdialogue/.

Franklin, Allan. 1994 "How to Avoid the Experimenter's Regress." *Studies in History and Philosophy of Science* 25: 463–91.

Friends of the Earth. 2013. *What we stand for.* Accessed March 28, 2013. http://www.foe.co.uk/what_we_do/about_us/friends_earth_values_beliefs.html.

Friends of the Earth Europe. 2004. "Throwing Caution to the Wind: A Review of the European Food Safety Authority and Its Work on Genetically Modified Foods and Crops." Accessed February 10, 2013. http://www.google.gr/url?sa=t&rct=j&q=&esrc=s&source=web&cd=1&cad=rja&ved=0CC8QFjAA&url=http%3A%2F%2Fwww.gmfreeireland.org%2Fresources%2Fdocuments%2FFOE%2FEFSAreport.pdf&ei=otEWUc-EJoestAaujYDYAg&usg=AFQjCNHnx3lcp683AP8Sq1feX-DdW-ONDg&sig2=b4hqKX09OWxESzm.

———. 2007. "Press." Friends of the Earth Europe. March. Accessed May 1, 2010. http://www.foeeurope.org/ GMOs/Index.htm .

Friends of the Earth International. 2010. "Reports." Friends of the Earth UK. February. Accessed May 20, 2010. <www.foe.co.uk/resource/reports/who_benefits_from_gm_crops.pdf>.

———. 2012. *A Wolf in Sheep's Clothing? An Analysis of the "Sustainable Intensification" of Agriculture.* Part of the Friends of the Earth International "Who Benefits" Series Investigating the Winners and Losers of Industrial Agriculture Models. Amsterdam: Friends of the Earth International.

Friesen, Lyle F., Alison G. Nelson, and Rene C. Van Acker. 2003 "Evidence of Contamination of Pedigreed Canola Brassica Napus Seedlots in Western Canada with Genetically Engineered Herbicide Resistance Traits." *Agronomy Journal* 95: 1342–47.

Garfinkel, Harold. 1967. *Studies in Ethnomethodology.* Englewood Cliffs, NJ: Prentice Hall.

Gee, David, and Andrew Stirling. 2004. "Late Lessons from Early Warnings: Improving Science and Governance under Uncertainty and Ignorance." In *The Precautionary Principle: Protecting Public Health, the Environment and the Future of Our Children,* edited by Marco Martuzzi and Joel A. Tickner, 93–120. World Health Organization.

Gee, David. 2013. "More or Less Precaution?" In *Late Lessons from Early Warnings: Science, Precaution, Innovation,* 643–69. European Environment Agency.

GeneWatch. 2012. *GM Crops and Foods in Britain and Europe.* Accessed December 3, 2012. http://www.genewatch.org/sub-568547.

———. 2013. *About GeneWatch.* Accessed March 28, 2013. http://www.genewatch.org/sub-396416.

GeneWatch UK. 2012. *Guide to EU Regulations.* Accessed November 27, 2012. http://www.genewatch.org/sub-530862.

Giddens, Anthony. 1979. *Central Problems in Social Theory: Action, Structure and Contradiction in Social Analysis.* London: Macmillan.

———. 1984. *The Constitution of Society.* Cambridge: Polity.

———. 1989. "A Reply to My Critics." In *Social Theory of Modern Societies: Anthony Giddens and His Critics*, edited by David Held and John B. Thompson, 249–301. Cambridge: Cambridge University Press.

———. 1993. *New Rules of Sociological Method.* 2nd ed. Stanford: Stanford University Press.

Gieryn, Thomas F. 1982. "Relativist/Constructivist Programmes in the Sociology of Science: Redundance and Retreat." *Social Studies of Science* 12, no. 2: 279–97.

Gingras, Yves, and Silvan S. Schweber. 1986. "Review: Constraints on Construction." *Social Studies of Science* 16, no. 2: 372–83.

Gingras, Yves. 1997. "The New Dialectics of Nature." *Social Studies of Science* 27, no. 2: 317–34.

———. 1999. "From the Heights of Metaphysics: A Reply to Pickering." *Social Studies of Science* 29, no. 2: 312–15.

Glover, Dominic. 2009. "Undying Promise: Agricultural Biotechnology's Pro-poor Narrative, Ten Years on." Working Paper 15. Brighton: STEPS Centre. Accessed December 12, 2011. http://www.steps-centre.org/PDFs/Bt%20Cotton%20web.pdf.

———. 2010. "Is Bt Cotton a Pro-poor technology? A Review and Critique of Empirical Record." *Journal of Agrarian Change* 10, no. 4: 482–509.

GM Freeze. 2012a. *Balancing Media Coverage.* Accessed December 3, 2012. http://www.gmfreeze.org/take-action/media-balance/.

———. 2012b. *Food Companies Reject GM Wheat.* Accessed December 4, 2012. http://www.gmfreeze.org/gmwheatnothanks/food-companies-reject-gm-wheat/.

———. 2012c. *GM Wheat? No Thanks!* Accessed November 21, 2012. http://www.gmfreeze.org/gmwheatnothanks/.

———. 2012d. *GM Wheat—What We Can Do.* Accessed November 27, 2012. http://www.gmfreeze.org/gmwheatnothanks/gm-wheat-what-we-can-do/.

———. 2013. *Join Us.* Accessed March 28, 2013. http://www.gmfreeze.org/join_us/.

GM Science Review Panel. 2003. *GM Science Review.* First Report. London: Department of Trade and Industry.

GMO Compass. 2006a. *Food and Feed from GMOs: The Long Road to Authorisation.* Accessed January 21, 2013. http://www.gmo-compass.org/eng/regulation/regulatory_process/157.eu_gmo_authorisation_procedures.html.

———. 2006b. *Freedom of Choice: Selecting the Deliberately Applied System.* December 21. Accessed January 29, 2016. http://www.gmo-compass.org/eng/regulation/coexistence/133.freedom_choice_selecting_deliberately_applied_system.html.

GMO Safety. 2012. *GM Wheat Debate in the UK.* Accessed November 24, 2012. http://www.gmo-safety.eu/news/1418.debate-genetically-modified-wheat-uk.html.

GMO-free Europe. 2005. *Berlin Manifesto for GMO-free Regions and Biodiversity in Europe.* January 23. Accessed January 31, 2016. http://www.gmo-free-regions.org/past-gmo-free-conferences/gmo-free-conference-2005/berlin-manifesto.html.

Godin, Benoît, and Yves Gingras. 2002. "The Experimenters' Regress: From Skepticism to Argumentation." *Studies in History and Philosophy of Science* 33: 137–52.

Goffman, Erving. 1959. *The Presentation of Self in Everyday Life*. New York: Anchor Books.

Goldthorpe, John. 2000. *On Sociology*. New York: Oxford University Press.

Goodman, Robert M. and Norman Newell. 1985. "Genetic Engineering of Plants for Herbicide Resistance: Status and Prospects." In *Engineered Organisms in the Environment: Scientific Issues*, edited by Harlyn O. Halvorson, David Pramer and Marvin Rogul, 47–53. Washington, DC: American Society for Microbiology.

Gough, Ian and Olofsson, G. (eds.). 2000. *Capitalism and Social Cohesion: Essays on Exclusion and Integration*. London: Macmillan.

Gouldner, Alvin W. 1976. *The Dialectic of Ideology and Technology*. New York: Seabury Press.

GRACE. 2012. *GRACE in Brief*. August 1. Accessed January 24, 2016. http://www.grace-fp7.eu/en/content/grace-brief.

———. 2015a. *Data Interpretation of an Anonymous Toxicologist and Corresponding GRACE Statements*. January 9. Accessed January 25, 2016. http://www.grace-fp7.eu/content/data-interpretation-anonymous-toxicologist-and-corresponding-grace-statements.

———. 2015b. *End of Discussion? Testbiotech Refuses to Join Public Scientific Debate*. January 14. Accessed January 24, 2016. http://www.grace-fp7.eu/en/content/end-discussion-testbiotech-refuses-join-public-scientific-debate.

Graham, Karen. 2016. *New Study by USDA Proves It Was Wrong about GE Alfalfa*. January 23. Accessed January 24, 2016. http://www.digitaljournal.com/news/environment/new-study-by-usda-proves-they-were-wrong-about-ge-alfalfa/article/455607.

Greenpeace. 2015. "Twenty Years of Failure: Why GM Crops Have Failed to Deliver on Their Promises." *Greenpeace International*. November 5. Accessed January 17, 2016. http://www.greenpeace.org/international/en/publications/Campaign-reports/Agriculture/Twenty-Years-of-Failure/.

Guardian. 2012. *GM Debate between Take the Flour Back and Rothamsted Research*. Accessed October 21, 2012. http://www.guardian.co.uk/environment/2012/jun/01/letter-take-flour-back-rothamsted?newsfeed=true.

Habermas, Jürgen. 1976. *Legitimation Crisis*. London: Heinemann.

Hall, Sarah. 1999. *Monsanto Ads Condemned*. March 1. Accessed February 7, 2016. https://www.theguardian.com/science/1999/mar/01/gm.food.

Hansen, Steffen Foss, and Joel A. Tickner. 2013. "The Precautionary Principle and False Alarms—Lessons Learned." In *Late Lessons from Early Warnings: Science, Precaution, Innovation*, 17–45. European Environment Agency.

Hargens, Lowell. 2004. "What is Mertonian Sociology of Science?" *Scientometrics* 60, no. 1: 63–70.

Harremoës, Poul, David Gee, Malcolm MacGavin, Andy Stirling, Jane Keys, Brian Wynne and Sofia Guedes Vaz. 2001. *Late Lessons from Early Warnings: The Precautionary Principle 1896–2000*. Copenhagen: European Environment Agency.

Heilbron, John L. 1986. "Review: Constructing Quarks: A Sociological History of Particle Physics." *The American Journal of Sociology* 91, no. 6: 1479–81.

Heilbroner, Robert. 1967. "Do Machines Make History?" *Technology and Culture* 8, no. 3: 335–45.

———. 1994. "Technological Determinism Revisited." In *Does Technology Drive History? The Dilemma of Technological Determinism*, edited by Merritt Smith and Leo Marx, 67–78. Cambridge, MA: The MIT Press.

Heinemann, Jack A., Melanie Massaro, Dorien S. Coray, Sarah Zanon Agapito-Tenfen and Jiajun Dale Wen. 2014. "Sustainability and Innovation in Staple Crop Production in the US Midwest." *International Journal of Agricultural Sustainability* 12, no. 1: 71–88.

Hertefeldt, Tina D., Rikke B. Jørgensen and Lars B. Pettersson. 2008. "Long-Term Persistence of GM Oilseed Rape in the Seedbank." *Biology Letters* 4, no. 3: 314–17.

Hesse, Mary. 1986. "Changing Concepts and Stable Order." *Social Studies of Science* 16: 714–26.

Hetherington, Kevin. 1999. "From Blindness to Blindness: Museums, Heterogeneity and the Subject." In *Actor Network Theory and After*, edited by John Law and John Hassard, 51–73. Oxford: Blackwell Publishers.

Heur, Bas van, Loet Leydesdorff and Sally Wyatt. 2012. "Turning to Ontology in STS? Turning to STS through 'Ontology.'" *Social Studies of Science* 43, no. 3: 341–62.

HGCA. *Exports and the UK Market*. 2012. Accessed December 4, 2012. http://www.hgca. com/content.output/5165/5165/Exports/Exports/Exports%20and%20the%20 UK%20market.mspx.

Higgins, Vaughan and Geoffrey Lawrence. *Agricultural Governance: Globalization and the New Politics of Regulation*. Abingdon: Routledge, 2009.

Hilbeck, Angelika et al. 2015. "No Scientific Consensus on GMO Safety." *Environmental Sciences Europe* 27, no. 4.

Holland, Nina. 2016. *Brussels Biotech Lobby's Last Push for "GM 2.0" Technologies to Escape Regulation*. February 2. Accessed February 16, 2016. http://www.theecologist.org/ News/news_analysis/2987034/brussels_biotech_lobbys_last_push_for_gm_20_tech- nologies_to_escape_regulation.html.

Horlick-Jones, Tom et al. 2007. *The GM Debate: Risk, Politics and Public Engagement*. London: Routledge.

House of Commons: Environment Committee. 1990. *The Environment White Paper: This Common Inheritance Cm 1200: Minutes of Evidence, Wednesday 21 November 1990: Department of the Environment*. London: H.M. Stationery Office. Accessed October 20, 2015. http:// data.parliament.uk/writtenevidence/committeeevidence.svc/evidencedocument/ science-and-technology-committee/gm-foods-and-application-of-the-precautionary- principle-in-europe/written/8640.pdf.

House of Commons Science and Technology Committee. 2014a. *Oral Evidence: GM Foods and Application of the Precautionary Principle in Europe, 29 October 2014*. HC 328. London: House of Commons.

———. 2014b. *Oral Evidence: GM Foods and Application of the Precautionary Principle in Europe, Monday 1 December 2014*. HC 328. London: House of Commons.

———. 2015a. *Advanced Genetic Techniques for Crop Improvement: Regulation, Risk and Precaution: Fifth Report of Session 2014–15*. Report Ordered by the House of Commons: HC 328. London: The Stationery Office Limited.

———. 2015b. *Advanced Genetic Techniques for Crop Improvement: Regulation, Risk and Precaution: Government Response to the Committee's Fifth Report of Session 2014– 15*. Third Special Report of Session 2015–16: HC 519. London: The Stationery Office Limited.

Howlett, Michael and David Laycock. 2012. *Regulating Next Generation Agri-Food Biotechnology: Lessons from European, North American, and Asian Experiences*. Abingdon: Routledge.

Hughes, Thomas. 1994. "Technological Momentum." In *Does Technology Drive History? The Dilemma of Technological Determinism*, edited by Merritt Smith and Leo Marx, 101–14. Cambridge, MA: The MIT Press.

IAASTD. 2009. *Agriculture at a Crossroads*. International Assessment of Agricultural Knowledge, Science and Technology for Development: Global Report. Washington, DC: Islands Press.

ILSI. 2004. "Nutritional and Safety Assessments of Foods and Feeds Nutritionally Improved through Biotechnology." Accessed February 17, 2013. www.ilsi.org/FoodBioTech/ Publications/02_Nutritional%20_Safety%20Assessment%20of%20GM%20Foods_ 2004.pdf.

IRGC. 2005. *An Introduction to the IRGC Risk Governance Framework.* Geneva: IRGC.

———. 2016a. *Mission and Purpose.* January 1. Accessed February 11, 2016. http://www. irgc.org/about/mission-and-purpose/.

———. 2016b. *IRGC Risk Governance Framework.* January 1. Accessed February 11, 2016. http://www.irgc.org/risk-governance/irgc-risk-governance-framework/.

IRRI. 2013. *Clarifying Recent News about Golden Rice.* February 21. Accessed January 14, 2016. http://irri.org/blogs/item/clarifying-recent-news-about-golden-rice.

———. 2014. *What is the Status of the Golden Rice Project Coordinated by IRRI?* March 1. Accessed January 14, 2016. http://irri.org/golden-rice/faqs/what-is-the-status-of-the-golden-rice-project-coordinated-by-irri.

Irwin, Alan. 2008. "STS Perspectives on Scientific Governance." In *The Handbook of Science and Technology Studies,* edited by Edward J. Hackett, Olga Amsterdamska, Michael Lynch and Judy Wajcman, 583–607. London: The MIT Press.

ISAAA. 2013. *Top Ten Facts about Biotech/GM Crops in 2012.* Accessed April 10, 2013. http:// www.isaaa.org/resources/publications/briefs/44/toptenfacts/default.asp.

Isidore, Chris. 2009. *General Motors Bankruptcy: End of an Era.* Accessed January 25, 2011. http://money.cnn.com/2009/06/01/news/companies/gm_bankruptcy/.

Jack, Lisa, and Ahmed Kholeif. 2007. "Introducing Strong Structuration Theory for Informing Qualitative Case Studies in Organization, Management and Accounting Research." *Qualitative Research in Organizations and Management: An International Journal* 2, no. 3: 208–25.

———. 2008. "Enterprise Resource Planning and a Contest to Limit the Role of Management Accountants: A Strong Structuration Perspective." *Accounting Forum* 32, no. 1: 30–45.

Jaffe, Adam B., and Josh Lerner. 2004. *Innovation and Its Discontents: How Our Broken Patent System is Endangering Innovation and Progress, and What to Do About It.* Princeton, NJ: Princeton University Press.

James Hutton Institute. 2014. *Written Evidence Submitted by the James Hutton Institute GMC0009.* May 7. Accessed October 11, 2015. http://data.parliament.uk/writtenevidence/ committeeevidence.svc/evidencedocument/science-and-technology-committee/gm-foods-and-application-of-the-precautionary-principle-in-europe/written/8133.pdf.

Jasanoff, Sheila. 1993 "Bridging the Two Cultures of Risk Analysis." *Risk Analysis* 13, no. 2: 123 29.

———. 2005. *Science and Democracy in Europe and the United States.* Princeton: Princeton University Press.

———. 2006a. "Ordering Knowledge, Ordering Society." In *States of Knowledge: The Co-production of Science and Social Order,* edited by Sheila Jasanoff, 13–45. Abingdon: Routledge.

———. 2006b. "The Idiom of Co-production." In *States of Knowledge: The Co-production of Science and Social Order,* edited by Sheila Jasanoff, 1–12. Abingdon: Routledge.

———. 2008. "Making Order: Law and Science in Action." In *The Handbook of Science and Technology Studies,* edited by Edward J. Hackett, Olga Amsterdamska, Michael Lynch and Judy Wajcman, 761–86. London: The MIT Press.

John Innes Centre. 2014. "Supplementary Written Evidence Submitted by John Innes Centre: GMC0058." *Commons Select Committee: GM foods and Application of the Precautionary*

Principle in Europe. December 1. Accessed November 1, 2015. http://data.parliament.uk/ writtenevidence/committeeevidence.svc/evidencedocument/science-and-technology-committee/gm-foods-and-application-of-the-precautionary-principle-in-europe/written/15723.pdf.

Kass, Gary. 2001. *Open Channels: Public Dialogue in Science and Technology.* POST. London: Parliament Office of Science and Technology.

Kaye-Blake, William H., Caroline M. Saunders and Selim Cagatay. 2008. "Genetic Modification Technology and Produce Returns: The Impacts of Productivity, Preferences, and Technology Uptake." Review of *Agricultural Economics* 30, 4: 692–710.

Keating, Dave. 2012. *Council Rejects EFSA Nominee.* Accessed February 10, 2013. http://www.europeanvoice.com/article/2012/june/council-rejects-efsa-nominee/74562.aspx.

Kinchy, Abby. 2012. *Seeds, Science, and Struggle: The Global Politics of Transgenic Crops.* London: The MIT Press.

Kleinman, Daniel Lee. 2003. *Impure Cultures—University Biology and the World of Commerce.* Wisconsin: The University of Wisconsin Press.

———. 2005. *Science and Technology in Society.* Oxford: Blackwell Publishing.

Kleinman, Daniel Lee, and Jack Kloppenburg Jr. 1991. "Aiming for the Discursive High Ground: Monsanto and the Biotechnology Controversy." *Sociological Forum* 6, no.3: 427–47.

Klintman, Mikael. 2002. "The Genetically Modified GM Food Labelling Controversy: Ideological and Epistemic Crossovers." *Social Studies of Science* 32, no. 1: 71–91.

Kloppenburg, Jack. 2010. "Impeding Dispossession, Enabling Repossession: Biological Open Source and the Recovery of Seed Sovereignty." *Journal of Agrarian Change* 10, no. 3: 367–88.

Knorr Cetina, Karin. 1982. "Scientific Communities of Transepistemic Arenas of Research? A Critique for Quasi-Economic Models of Science." *Social Studies of Science* 12: 101–30.

———. 1991. "Merton's Sociology of Science: The First and the Last Sociology of Science?" *Contemporary Sociology* 20, no. 4 July: 522–26.

———. 1998. "Citation for H. M. Collins." *Science, Technology, and Human Values* 23, no. 4: 491–93.

Kok, E. J., and Harry A. Kuiper. 2003. "Comparative Safety Assessment for Biotech Crops." *Trends in Biotechnology* 21: 439–44.

Kriebel, David, Joel A. Tickner, Paul Epstein, John Lemons, Richard Levins, Edward L. Loechler, Margaret Quinn, Ruthann Rudel, Ted Schettler and Michael Stoto. 2004. "The Precautionary Principle in Environmental Science." In *The Precautionary Principle: Protecting Public Health, the Environment and the Future of Our Children,* edited by Marco Martuzzi and Joel A. Tickner, 145–65. World Health Organization. Access December 5, 2015. http://www.euro.who.int/en/publications/abstracts/precautionary-principle-the-protecting-public-health,-the-environment-and-the-future-of-our-children.

Krimsky, Sheldon. 2015. "An Illusory Consensus behind GMO Health Assessment." *Science, Technology, and Human Values* 40, no. 6: 883–914.

Kunert, Grit, Carolina Reinhold and Jonathan Gershenzon. 2010. "Constitutive Emission of the Aphid Alarm Pheromone, E-β-farnesene, from Plants Does Not Serve as a Direct Defense against Aphids." *BMC Ecology* 10, no. 23: 1–12.

Larrain, Jorge. 1979. *The Concept of Ideology.* London: Hutchinson.

Latour, Bruno. 1988a. "A Relativistic Account of Einstein's Relativity." *Social Studies of Science* 18, no. 1: 3–44.

———. 1988b. *The Pasteurization of France.* Cambridge, MA: Harvard University Press.

———. 1994 "Pragmatogonies: 'A Mythical Account of How Humans and Nonhumans Swap Properties.'" *American Behavioral Scientist* 37, no. 6: 791–808.

———. 1996. "On Actor-Network Theory. A Few Clarifications Plus More than a few Complications." *Bruno Latour Website.* Accessed July 21, 2013. www.bruno-latour.fr/ sites/default/files/P-67%20ACTOR-NETWORK.pdf

———. 1997. "The Trouble with Actor-Network Theory." *Latour-Clarifications.* Accessed January 27, 2011. http://www.cours.fse.ulaval.ca/edc-65804/latour-clarifications. pdf.

———. 1999. "On Recalling ANT." In *Actor-Network Theory and After,* edited by John Law and John Hassard, 15–25. Oxford: Blackwell Publishing.

———. 2005. *Reassembling the Social: An Introduction to Actor-Network-Theory.* Oxford: Oxford University Press.

———. 2010. *Bruno Latour.* Accessed October 29, 2010. http://www.bruno-latour.fr/ faq-en.html.

Latour, Bruno, and Steve Woolgar. 1986. *Laboratory Life: The Construction of Scientific Facts.* 2nd ed. Princeton: Princeton University Press.

Laudan, Larry. 1982. "A Note on Collins' Blend of Relativism and Empiricism." *Social Studies of Science* 12, no. 1: 131–32.

Law, John. 1992. "Notes on the Theory of the Actor-Network: Ordering, Strategy and Heterogeneity." *Systems Practice* 5: 379–93.

———. 1997. *Traduction/Trahison: Notes on ANT.* Accessed August 1, 2013. http://cseweb. ucsd.edu/~goguen/courses/175/stslaw.html.

———. 1999. "After ANT: complexity, naming and topology." In *Actor-Network Theory and After,* edited by John Law and John Hassard, 1–14. Oxford: Blackwell Publishers.

———. 2002. *Aircraft Stories: Decentering the Subject in Technoscience.* Durham, NC: Duke University Press.

———. 2013. *Prof John Law.* Accessed July 27, 2013. http://www.open.ac.uk/ socialsciences/staff/people-profile.php?name=John_Law.

Law, John and John Hassard. 1999. *Actor-Network Theory and After.* Oxford: Blackwell.

Law, John and Michael Callon. 1992. "The Life and Death of an Aircraft: A Network Analysis of Technical Change. In *Shaping technology / Building Societies: Studies in Sociotechnical Change,* edited by Wiebe Bijker and John Law, 21–52. Baskerville: MIT Press.

Layder, Derek. 2006. *Understanding Social Theory.* 2nd ed. London: Sage Publications.

Lee, Maria. 2008. *EU Regulation of GMOs.* Cheltenham: Edward Elgar Publishing.

Levidow, Les. 2009. "Making Europe Unsafe for Agbiotech." In *Handbook of Genetics and Society,* edited by Paul Atkinson, Peter Glasner and Margaret Lock, 110–26. London: Routledge.

Levidow, Les and Susan Carr. 2010. *GM Food on Trial: Testing European Democracy.* London: Routledge.

Levidow, Les and Joseph Murphy. 2006. *Governing the Transatlantic Conflict over Agricultural Biotechnology.* London: Routledge.

Lezaun, Javier. 2006. "Creating a New Object of Government: Making Genetically Modified Organisms Traceable." *Social Studies of Science* 36, no. 4: 499–531.

Lockwood, David. 1964. "Social Integration and System Integration." In *Solidarity and Schism,* by David Lockwood, 399–412. Oxford: Oxford University Press.

————. 1992. "Social Integration and System Integration." In *Solidarity and Schism: "The Problem of Disorder" in Durkheimian and Marxist Sociology*, 399–412. Oxford: Clarendon Press.

Lomborg, Björn. 2013. *A Golden Rice Opportunity*. February 15, Accessed January 14, 2016. http://www.project-syndicate.org/commentary/the-costs-of-opposing-gm-foods-by-bj-rn-lomborg.

Lorch, Antje. 2010. *Amflora Authorized for Contamination*. Accessed March 29, 2013. http://www.ifrik.org/blog/amflora-authorized-contamination.

Losey, John E., Linda S. Rayor and Maureen E. Carter. 1999. "Transgenic Pollen Harms Monarch Larvae." *Nature* 399: 6–7.

Luker, Sara. 2012. *Rothamsted Research Praised for Comms against GM Protest Action*. Accessed December 1, 2012. http://www.prweek.com/uk/news/1133950/rothamsted-research-praised-comms-against-gm-protest-action/.

Lundgren, Peter. 2012. *GM Wheat in the UK? No Thanks!* Accessed December 4, 2012. http://www.peterlundgren.co.uk/2012/03/03/gm-wheat-in-the-uk-no-thanks-6/

Lynas, Mark. 2015. *With G.M.O. Policies, Europe Turns Against Science*. October 24. Accessed January 24, 2016. http://www.nytimes.com/2015/10/25/opinion/sunday/with-gmo-policies-europe-turns-against-science.html.

Lynch, Michael. 1996. "Review: The Mangle of Practice: Time, Agency, and Science by Andrew Pickering." *Contemporary Sociology* 25, no. 6: 809–11.

Ma, B. L. and K. D. Subedi. 2005. "Development, Yield, Grain Moisture, and Nitrogen Uptake of Bt Corn Hybrids and Their Conventional Near-Isolines." *Field Crops Research* 93: 199–211.

MacKenzie, Donald. 1984. "Marx and the Machine." *Technology and Culture* 25, no. 3: 473–502.

MacKenzie, Donald, and Judy Wajcman. 1999. *The Social Shaping of Technology*. 2nd ed. Berkshire: Open University Press.

Malkan, Stacy. 2016. "Why Is Cornell University Hosting a GMO Propaganda campaign?" January 22. Accessed January 24, 2016. http://www.theecologist.org/News/news_analysis/2986952/why_is_cornell_university_hosting_a_gmo_propaganda_campaign.html.

Marx, Karl. 1993. *Grundrisse*. London: Penguin Books.

————. 1998. *The German Ideology*. New York: Prometheus Books.

————. 2001. *The Poverty of Philosophy*. Chicago: Elibron Classics.

Matthewman, Steve. 2011. *Technology and Social Theory*. London: Palgrave Macmillan.

Matthews, Jonathan. 2012. "Science One, Whining Greenies Nil." Accessed November 30, 2012. http://www.spinwatch.org.uk/-articles-by-category-mainmenu-8/46-gm-industry/5505-science-one-whining-greenies-nil.

McKie, Robin. 2013. "After 30 Years, is a GM Food Breakthrough Finally Here?" February 2. Accessed January 14, 2016. http://www.theguardian.com/environment/2013/feb/02/genetic-modification-breakthrough-golden-rice.

McLellan, David. 1995. *Concepts in Social Thought: Ideology*. 2nd ed. Minneapolis: University of Minnesota Press.

Merton, Robert K. 1936. "The Unanticipated Consequences of Social Action." In *On Social Structure and Science*, 173–82. Chicago: University of Chicago Press.

————. 1938. "Science and the Social Order." *Philosophy of Sciencebenton* 5, no. 3: 321–37.

————. 1942. "The Normative Structure of Science." In *The Sociology of Science*, by Robert K. Merton, 267–280. Chicago: University of Chicago Press.

————. 1952. "The Neglect of the Sociology of Science." In *The Sociology of Science*, by Robert K. Merton, 210–22. Chicago: University of Chicago Press.

————. 1973. *The Sociology of Science: Theoretical and Empirical Investigations.* Edited by Norman W Storer. Chicago: University of Chicago Press.

Messean, Antoine, Frédérique Angevin, Manuel Gómez-Barbero, Klaus Menrad and Emilio Rodríguez-Cerezo. 2006. "New Case Studies on the Co-existence of GM and Non-GM Crops in European Agriculture." *Joint Research Center.* IPTS Publications. Accessed June 14, 2013. http://ftp.jrc.es/EURdoc/22102-ExeSumm.pdf.

Michael, Mike. 2000. *Reconnecting Culture, Technology, and Nature.* London: Routledge.

Michalopoulos, Sarantis. 2015. *New Plant Breeding Techniques: Innovation Breakthrough or GMOs in Disguise?* October 22. Acccssed February 16, 2016. http://www.euractiv.com/section/science-policymaking/news/new-plant-breeding-techniques-innovation-breakthrough-or-gmos-in-disguise/.

————. 2016. *Commission to Decide on New Plant Breeding Techniques within Three Months.* January 8. Accessed February 16, 2016. http://www.euractiv.com/section/agriculture-food/news/commission-to-decide-on-new-plant-breeding-techniques-within-three-months/.

Mills, Charles W. 1959. *The Sociological Imagination.* Oxford: Oxford University Press.

Misa, Thomas J. 1988. "How Machines Make History, and How Historians and Others Help Them to DoSso." *Science, Technology and Human Values* 13, no. 3/4: 308–31.

————. 1992a. "Controversy and Closure in Technological Change: Constructing 'Steel'." In *Shaping Technology/Building Society,* edited by Wiebe Bijker and John Law, 109–39. Cambridge, MA: The MIT Press.

————. 1992b. "Theories of Technological Change: Parameters and Purposes." *Science, Technology, and Human Values* 17, no. 1: 3–12.

————. 1994. "Retrieving Sociotechnical Change from Technological Determinism." In *Does Technology Drive History? The Dilemma of Technological Determinism,* edited by Merritt Smith and Leo Marx, 115–42. Cambridge, MA: The MIT Press.

Moloney, Maurice. 1995. *Environmental Advantages from Plant Genetic Engineering.* Accessed November 27, 2012. http://people.ucalgary.ca/~pubconf/Education/moloney.htm

Mouzelis, Nicos. 1967. *Organisation and Bureaucracy: An Analysis of Modern Theories.* Chicago: Kegan Paul.

————. 1979.*Modern Greece: Facets of Underdevelopment.* 2nd ed. London: Macmillan.

————. 1986. *Politics in the Semi-periphery: Early Parliamentarism and Late Industrialisation in the Balkans and Latin America.* London: Macmillan.

————. 1990. *Post-Marxist Alternatives: The Construction of Social Orders.* London: Macmillan Press.

————. 1992. "The Interaction Order and the Micro-Macro Distinction." *Sociological Theory* 10, no. 1: 122–28.

————. 1994. *Back to Sociological Theory: The Construction of Cocial Orders.* New York: St. Martin's Press.

————. 1995. *Sociological Theory: What Went Wrong?* London: Routledge.

————. 2008.*Modern and Postmodern Social Theorizing: Bridging the Divide.* Cambridge: Cambridge University Press.

Müller, Werner. 2003. *Concepts for Coexistence.* ECO-RISK: Office of Ecological Risk Research, Vienna: Commissioned by the Federal Ministry of Health and Women.

Nandula, Vijay, Reddy Krishna, Duke Stephen and Daniel Poston. 2005. "Glyphosate-resistant Weeds: Current Status and Future Outlook." *Outlooks on Pest Management*: 183–87.

National Centre for Biotechnology Education. 2006. *UK National Consensus Conference on Plant Biotechnology 1994.* Accessed November 26, 2015. http://www.ncbe.reading.ac.uk/NCBE/GMFood/conference.html.

National Farmers' Union. 2014. *Written Evidence Submitted by the National Farmers' Union NFU GM0022.* May 7. Accessed November 2, 2015. http://data.parliament.uk/writtenevidence/committeeevidence.svc/evidencedocument/science-and-technology-committee/gm-foods-and-application-of-the-precautionary-principle-in-europe/written/8609.pdf.

Nelsen, Arthur. 2015. *Half of Europe Opts Out of New GM Crop Scheme.* October 1. Accessed November 2, 2015. http://www.theguardian.com/environment/2015/oct/01/half-of-europe-opts-out-of-new-gm-crop-scheme.

Nelsen, Arthur. 2014. *NGO Backlash to Chief Scientific Advisor Position Grows.* August 19. Accessed October 22, 2015. http://www.euractiv.com/sections/science-policymaking/ngo-backlash-chief-scientific-advisor-position-grows-307823.

Newsnight. 2012. *GM Debate—Newsnight.* Accessed December 1, 2012. http://www.youtube.com/watch?v=EfFszpXV8x0.

Newton-Smith, W. H. 1981. *The Rationality of Science.* London: Routledge & Kegan Paul Ltd.

Nightingale, Paul. 2014. "Written Evidence Submitted by Professor Paul Nightingale GMC0045." *Commons Select Committee: GM Foods and Application of the Precautionary Principle in Europe.* May 7. Accessed December 2, 2015. http://data.parliament.uk/writtenevidence/committeeevidence.svc/evidencedocument/science-and-technology-committee/gm-foods-and-application-of-the-precautionary-principle-in-europe/written/8787.pdf.

Northern Ireland Executive. 2015. *Durkan Bans GM Crops in Northern Ireland.* September 21. Accessed November 7, 2015. http://www.northernireland.gov.uk/news-doe-210915-durkan-bans-gm.

O' Riordan, Timothy, and Andrew Jordan. 1995. "The Precautionary Principle, Science, Politics and Ethics." *Centre for Social and Economic Research on the Global Environment.* Accessed May 4, 2013. http://www.cserge.ac.uk/sites/default/files/pa_1995_02.pdf.

Oke, Krista B., Peter A. H. Westley, Darek T. R. Moreau and Ian A. Fleming. 2013. "Hybridization between Genetically Modified Atlantic Salmon and Wild Brown Trout Reveals Novel Ecological Interactions." *Proceedings of the Royal Society B: Biological Sciences* 280, 1763. doi:10.1098/rspb.2013.1047.

Olson, Dennis. 2005. "Hard Red Spring Wheat at a Genetic Crossroad: Rural Prosperity or Corporate Hegemony?" In *Controversies in Science and Technology: From Maize to Menopause,* edited by Daniel Lee Kleinman, Abby J. Kinchy and Jo Handelsman, 150–68. Wisconsin: The University of Wisconsin Press.

Olson, Mancur. 1998. *The Logic of Collective Action.* London: Harvard University Press.

Ostrower, Jon, and Andy Pasztor. 2013. *Dreamliner's Other Issues Draw Attention.* Accessed June 6, 2013. http://online.wsj.com/article/SB10001424127887323463704578493442061830754.html.

Outhwaite, William, ed. 1993. *The Blackwell Dictionary of Modern Social Thought.* Oxford: Blackwell Publishing.

Ozawa, Connie P. 1991. *Recasting Science: Consensual Procedures in Public Policy Making.* Oxford: Westview Press.

———. 1996. "Science in Environmental Conflicts." *Sociological Perspectives* 39, no. 2: 219–30.

Ozawa, Connie P., and Lawrence Susskind. 1985 "Mediating Science-Intensive Policy Disputes." *Journal of Public Analysis and Management* 5, no. 1: 23–39.

Paine, Jacqueline A. et al. 2005. "Improving the Nutritional Value of Golden Rice through Increased Pro-vitamin A Content." *Nature Biotechnology* 23, no. 4: 482–87.

Park, Julian, Ian McFarlane, Richard Phipps and Graziano Ceddia. 2011 "The Impact of the EU Regulatory Constraint of Transgenic Crops on Farm Income." *New Biotechnology* 28, no. 4: 396–406.

Patterson, Owen. 2013. *Owen Paterson Speech at the National Farmers Union Annual Conference 2013.* February 27. Accessed November 6, 2015. https://www.gov.uk/government/speeches/owen-paterson-speech-at-the-national-farmers-union-annual-conference-2013.

Paul, Helena, Ricarda Steinbrecher, Lucy Michaels and Devlin Kuyek. 2003. *Hungry Corporations: Transnational Biotech Companies Colonise the Food Chain.* London: Zed Books.

Perry, Joe N. 2010. "A Mathematical Model of Exposure of Non-target Lepidoptera to Bt-Maize Pollen Expressing Cry1Ab." *ISB News Report.* May 1. Accessed February 8, 2016. http://www.isb.vt.edu/news/2010/May10.pdf.

———. 2014. "Written Evidence Submitted by Professor Joe N. Perry GMC0016." *Commons Select Committee: GM Foods and Application of the Precautionary Principle in Europe.* October 22. Accessed October 29, 2015. http://data.parliament.uk/writtenevidence/committeeevidence.svc/evidencedocument/science-and-technology-committee/gm-foods-and-application-of-the-precautionary-principle-in-europe/written/14811.pdf.

Perry, Joe N. et al. 2010. "A Mathematical Model of Exposure of Nontarget Lepidoptera to Bt-maize Pollen Expressing Cry1Ab within Europe." *Proceedings of the Royal Society B* 277: 1417–25.

Pickering, Andrew. 1981. "Constraints on Controversy: The Case of the Magnetic Monopole." *Social Studies of Science* 11, no. 1: Special Issue: *Knowledge and Controversy: Studies of Modern Natural Science* 69–93.

———. 1993. "The Mangle of Practice: Agency and Emergence in the Sociology of Science." *American Journal of Science* 99, no. 3: 559–89.

———. 1995. *The Mangle of Practice: Time, Agency, and Science.* Chicago: The University of Chicago Press.

Pilson, Diana, and Holly R. Prendeville. 2004 "Ecological Effects of Transgenic Crops and the Escape of Transgenes into Wild Populations." *Annual Review of Ecology, Evolution, and Systematics* 35: 149–74.

Pinch, T., and W. Bijker. 1984. "The Social Construction of Facts and Artefacts: Or How the Sociology of Science and the Sociology of Technology Might Benefit Each Other." *Social Studies of Science* 14: 399–441.

Ponti, Luigi. 2005. "Transgenic Crops and Sustainable Agriculture in the European Context." *Bulletin of Science, Technology & Society* 25, no. 4: 289–305.

Purvis, Trevor, and Alan Hunt. 1993. "Discourse, Ideology, Discourse, Ideology, Discourse, Ideology [...]" *British Journal of Sociology* 44, no. 3: 473–99.

Raffensperger, Carolyn, and Joel A. Tickner. 1999. *Protecting Public Health and the Environment: Implementing the Precautionary Principle.* Washington, DC: Island Press.

Ramsay, Gavin, Caroline Thompson and Geoff Squire. 2003. *Quantifying Landscape-Scale Gene Flow in Oilseed Rape.* Final Report of DEFRA Project RG0216, DEFRA & Scottish Research Institute, Dundee: Scottish Crop Research Institute.

Renevier, Laurent, and Mark Henderson. 2002. "Science and Scientists in International Environmental Negotiations." In *Transboundary Environmental Negotiation: New Approaches to Global Cooperation,* edited by Lawrence Susskind, William Monnmaw and Kevin Gallagher, 107–29. San Francisco: Jossey-Bass.

Riley, Pete. 2012. "Publications." *GM Freeze.* GM Freeze. Accessed December 2, 2012. http://www.gmfreeze.org/site_media/uploads/publications/gmwheataugust_2011final.pdf.

Ritzer, George. 2007. "Reviewed Works: Structuration Theory by Rob Stones." *Contemporary Sociology* 36, no. 1: 84–85.

Roth, Wolff Michael. 1998. *Designing Communities.* Dordrecht: Kluwer Academic Publishers.

Rothamsted Research. 2008. "Award Details." *BBSRC.* Accessed October 11, 2012. http://www.bbsrc.ac.uk/pa/grants/AwardDetails.aspx?FundingReference=BB/G004781/1.

———. 2011a. "APPLICATION FOR CONSENT TO RELEASE A GMO—HIGHER PLANTS." *Defra—Department for Environment, Food, and Rural Affairs.* Accessed June 15, 2012. http://archive.defra.gov.uk/environment/quality/gm/regulation/documents/11-r8-01-app-a.pdf.

———. 2011b. "PART B: INFORMATION ABOUT THE RELEASE APPLICATION TO BE INCLUDED ON THE PUBLIC REGISTER." *Defra—Department for Environment, Food, and Rural Affairs.* Accessed June 15, 2012. http://archive.defra.gov.uk/environment/quality/gm/regulation/documents/11-r8-01-app-b.pdf.

———. 2012a. *Collaborations: Rothamsted Research.* Accessed September 11, 2012. http://www.rothamsted.ac.uk/kec/Collaborations.php.

———. 2012b. *Corporate Information.* Accessed September 9, 2012. http://www.rothamsted.ac.uk/Content.php?Section=AboutUs&Page=CorporateInformation.

———. 2012c. *People: Professor Maurice Moloney.* Accessed September 11, 2012. http://www.rothamsted.ac.uk/PersonDetails.php?Who=1086

———. 2012d. "Rothamsted Research: Scientific Publications for scientists." Rothamsted Research. Accessed September 6, 2012. http://www.rothamsted.ac.uk/Content/pdfs/RRes_Strategy.pdf.

———. 2012e. *Rothamsted Wheat Trial.* Accessed November 8, 2012. http://www.rothamsted.ac.uk/Content.php?Section=AphidWheat&Page=QA.

———. 2012f. *Rothamsted Wheat Trial: Second Generation GM Technology to Emulate Natural Plant Defence Mechanisms.* Accessed December 3, 2012. http://www.rothamsted.ac.uk/Content.php?Section=AphidWheat.

———. 2012g. "Science Strategy 2012–2017." Accessed April 1, 2013. http://www.rothamsted.ac.uk/Content/pdfs/RRes_Strategy.pdf.

———. 2012h. *Youtube: Rothamsted Research.* Accessed September 2, 2012. http://www.youtube.com/user/RothamstedResearch.

———. 2013. *Professor Maurice Moloney to Join the Commonwealth Scientific and Industrial Research Organisation CSIRO—See more at: http://www.rothamsted.ac.uk/news/professor-maurice-moloney-join-commonwealth-scientific-and-industrial-research-organisation#sthash.j.* August 8. Accessed December 17, 2015. http://www.rothamsted.ac.uk/news/professor-maurice-moloney-join-commonwealth-scientific-and-industrial-research-organisation.

———. 2015. *Scientists Disappointed at Results from GM Wheat Field Trial.* June 25. Accessed December 3, 2015. http://www.rothamsted.ac.uk/news-views/scientists-disappointed-results-gm-wheat-field-trial.

Rowell, Andrew. 2003. *Don't Worry: It's Safe to Eat.* London: Earthscan Publications.

Rudebeck, Lars, Olle Törnquist and Virgilio Rojas. 1998. *Modernity, Late Development and Civil Society.* London: Macmillan.

Sargent, Rose-Mary. 1998. "Reviewed Work: The Mangle of Practice: Time, Agency, and Science by Andrew Pickering." *Philosophy of Science* 65, no. 4: 721–22.

Sarich, Christina. 2015. *Wales Announces Complete Ban on GMOs with 15 Other EU Countries.* October 5. Accessed November 7, 2015. http://www.globalresearch.ca/wales-announces-complete-ban-on-gmos-with-15-other-eu-countries/5479845.

Saunders, Harold H. 2001. *A Public Peace Process: Sustained Dialogue to Transform Racial and Ethnic Conflicts*. Vol. 9. Basingstoke: Palgrave.

Saunders, Peter. 2010. "The Precautionary Principle." In *Policy Responses to Societal Concerns in Food and Agriculture: Proceedings of an OECD Workshop*, 47–58. OECD. Accessed October 8, 2015. http://www.oecd.org/agriculture/agricultural-policies/policyresponsestoso-cietalconcernsinfoodandagricultureproceedingsofanoecdworkshop.htm.

Sayer, Derek. 1987. *The Violence of Abstraction: The Analytic Foundations of Historical Materialism*. Oxford: Blackwell Publishing.

Schatzberg, Eric. (1999). *Wings of Wood, Wings of Metal*. Princeton, NJ: Princeton University Press.

Schettler, Ted, and Carolyn Raffensperger. 2004. "Why Is a Precautionary Approach Needed?" In *The Precautionary Principle: Protecting Public Health, the Environment and the Future of Our Children*, edited by Marco Martuzzi and Joel A. Tickner, 63–83. World Health Organization.

Schiemann, Joachim. 2014. "Open Letter to Testbiotech in Response to Its Report and Press Release Dated 7-11-2014." *GRACE*. November 10. Accessed January 24, 2016. http://www.grace-fp7.eu/sites/default/files/GRACE%20Answer2TestBiotech_final_0.pdf

Schurman, Rachel and William A. Munro. 2010. *Fighting for the Future of Food: Activists versus Agribusiness in the Struggle over Biotechnology*. Minneapolis: University of Minnesota Press.

Sciencewise. 2011. "Talking about GM: Approaches to Public and Stakeholder Engagement." A paper by the Sciencewise-ERC subgroup on GM dialogue September 2011, London: Sciencewise.

Scott, John. 1996. "Reviewed Works: Sociological Theory: What Went Wrong? by Nicos Mouzelis." *The British Journal of Sociology* 47, no. 4: 731–32.

Sears, Mark K. et al. 2001. "Impact of Bt Corn Pollen on Monarch Butterfly Populations: A Risk Assessment." *Proceedings of the National Academy of Sciences of the United States of America* 98, no. 21: 11937–42.

Secretariat of the Convention on Biological Diversity. 2000. "Text of the Cartagena Protocol on Biosafety." Convention on Biological Diversity. Accessed May 7, 2013. http://bch.cbd.int/database/attachment/?id=10694.

Seed.Feed.Food. 2013. "Science Not Fiction: Time to Think Again about GM." Accessed April 10, 2013. http://seedfeedfood.eu/wp-content/uploads/2013/02/flipbook2.pdf.

Sense About Science. 2012. "Plant Scientists Answer Your Questions." Accessed September 26, 2012. http://www.senseaboutscience.org/pages/plant-science-qa.html.

———. 2015. "To the Cabinet Secretary for Rural Affairs and the Environment in the Scottish Government." Sense about Science. August 17. Accessed January 23, 2016. https://www.google.gr/url?sa=t&rct=j&q=&esrc=s&source=web&cd=6&cad=rja&uact=8&ved=0ahUKEwjtjdra3MDKAhWBNxQKHdJvBqEQFgg7MAU&url=http%3A%2F%2Fwww.senseaboutscience.org%2Fdata%2Ffiles%2FGM%2FLetter_to_Mr_Lochhead_17_Aug_2015.pdf&usg=AFQjCNGbZXeux81w4fTfE0ul8.

Séralini, Gilles-Eric, Dominique Cellier and Joël S. de Vendomois. 2007. "New Analysis of a Rat Feeding with a Genetically Modified Maize Reveals Signs of Hepatorenal Toxicity." *Archives of Environmental Contamination and Toxicology* 52, no. 4: 596–602.

Séralini, Gilles-Eric et al. 2012. "Long-term Toxicity of a Roundup Herbicide and a Roundup-tolerant Genetically odified Maize." *Food and Chemical Toxicology* 50: 4221–31.

Shapin, Steven. 1995. "Here and Everywhere: Sociology of Scientific Knowledge." *Annual Review of Sociology* 21: 289–321.

Shaw, William H. 1979. "'The Handmill Gives You the Feudal Lord': Marx's Technological Determinism." *History and Theory* 18, no. 2: 155–76.

Sibeon, Roger. 2004. *Rethinking Social Theory*. London: Sage Publications.

Sivin, Nathan. 1991. "Science, Religion, and Boundary Maintenance." *Contemporary Sociology* 20, no. 4: 526–30.

Sjöstedt, Gunnar, ed. 1993. *International Environmental Negotiation*. San Francisco: Sage Publications.

———. 2009. "Resolving Ecological Conflicts: Typical and Special Circumstances." In *The Sage Handbook of Conflict Resolution*, edited by Jacob Bercovitch, Victor Kremenyuk and William Zartman, 225–45. London: Sage.

Smith, Merritt, and Leo Marx. 1994. *Does Technology Drive History? The Dilemma of Technological Determinism*. Cambridge, MA: The MIT Press.

Soffritti, Morando, Fiorella Belpoggi, Eva Tibaldi, Davide Degli Esposti and Michelina Lauriola. 2007. "Life-span Exposure to Low Doses of Aspartame Beginning During Prenatal Life Increases Cancer Effects in Rats." *Environmental Health Perspectives* 115, no. 9: 1293–97.

Späth, Beat. 2015. *Politics vs. Science*. April 22. Accessed February 7, 2016. http://european-seed.com/politics-vs-science/.

Steinbrecher, Ricarda A. 2011. "GE Wheat Trial by Rothamsted Research." *EcoNexus*. Accessed September 13, 2012. http://www.econexus.info/sites/econexus/files/R.Steinbrecher%20-%20Econexus%20submission%20re%20GM%20Wheat%20application.pdf

STEPS Centre. 2014. "Written Evidence Submitted by Steps Centre GMC0004." *Commons Select Committee: GM Foods and Application of the Precautionary Principle in Europe*. May 7. Accessed November 15, 2015. http://data.parliament.uk/writtenevidence/committeeevidence.svc/evidencedocument/science-and-technology-committee/gm-food-and-application-of-the-precautionary-principle-in-europe/written/7134.pdf.

Stirling, Andy. 2013. "The Precautionary Principle." In *A Companion to the Philosophy of Technology*, edited by Jan Kyrre Berg Olsen Friis, Stig Andur Pedersen and Vincent F. Hendricks, 248–62. Chichester: Wiley-Blackwell.

———. 2014. "Making Choices in the Face of Uncertainty: Strengthening Innovation Democracy." In *Annual Report of the Government Chief Scientific Adviser 2014. Innovation: Managing Risk, Not Avoiding It. Evidence and Case Studies*, edited by Government Chief Scientific Adviser, 49–62. Government Office for Science.

Stones, Rob. 1996. *Sociological Reasoning: Towards a Past-modern Sociology*. Basingstoke: Macmillan Press.

———. 2005. *Structuration Theory*. London: Palgrave Macmillan.

Stones, Rob, and Sandra Moog. 2009. "Introduction: Intricate Webs—Nature, Social Relations, and Human Needs in the Writings of Ted Benton." In *Nature, Social Relations and Human Needs: Essays in Honour of Ted Benton*, edited by Sandra Moog and Rob Stones, 1–43. Basingstoke: Palgrave Macmillan.

Stones, Rob, and Trisha Greenhalgh. 2010. "Theorizing Big IT Programs in Healthcare: Strong Structuration Theory Meets Actor-Network Theory." *Social Science and Medicine* 70: 1285–94.

Storey, John. 2006. *Cultural Theory and Popular Culture: An Introduction*. 4th ed. Harlow: Pearson Education Limited.

Strategy Unit. 2003. *Field Work: Weighing Up the Costs and Benefits of GM Crops*. London: Cabinet Office.

Susskind, Lawrence. 1994. *Environmental Diplomacy: Negotiating More Effective Global Agreements.* New York: Oxford University Press.

Susskind, Lawrence, William Moomaw and Kevin Gallagher. 2002. *Transboundary Environmental Negotiation: New Approaches to Global Cooperation.* San Francisco: Jossey-Bass.

Svensson, Frances. 1979. "The Technological Challenge to Political Theory." In *The History and Philosophy of Technology*, edited by G. Bugliarello and D. B. Doner, 294–308. Urbana: University of Illinois Press.

Syngenta. 2014. "Written Evidence Submitted by Syngenta GM0036." *Commons Select Committee: GM Foods and Application of the Precautionary Principle in Europe.* May 7.

Tait, Joyce. 2012. "Risk Governance of Genetically Modified Crops—European and American Perspectives." *Risk Governance Framework.* April 1. Accessed February 11, 2016. http://irgc.org/wp-content/uploads/2012/04/Chapter_7_GM_Crops_final.pdf.

Take the Flour Back. 2012. *Why a Decontamination.* Accessed December 1, 2012. http://taketheflourback.org/why-a-decontamination/.

Take The Flour Back. 2012a. *An Open Letter to Rothamsted.* Accessed 10 September 2012. .http://taketheflourback.org/open-letter-to-rothamsted/

Take the Flour Back, 2012b. *The Wheat Trial.* Accessed November 24, 2012. http://taketheflourback.org/the-wheat-trial/

Taverne, Dick. 2004. "Let's Be Sensible about Public Participation." *Nature* 432: 271.

Testbiotech. 2012. "A Playground of the Biotech Industry?" *Testbiotech e.V.—Institute for Independent Impact Assessment in Biotechnology.* Accessed February 8, 2013. http://www.testbiotech.de/sites/default/files/TBT%20Background%20on%20EFSA_Conflict%20of%20Interests.pdf.

———. 2013. *Christoph Then.* Accessed March 29, 2013. http://www.testbiotech.de/en/user/6.

The European Union Center of North Carolina. 2007. "The EU-US Dispute over GMOs: Risk Perceptions and the Quest for Regulatory Dominance." University of North Carolina. Accessed December 6, 2012. http://www.unc.edu/depts/europe/business_media/businessbriefs/Brief0705-GMOs.pdf.

The Future of Food. 2007. DVD. Directed by Deborah Koons. Produced by Arts Alliance Amer.

The Greens in the European Parliament. 2010. *EFSA and GMOs.* Accessed February 10, 2013. http://www.greens-efa.eu/efsa-and-gmos-2628.html.

The Machine That Changed the World. 1992. TV miniseries. Directed by Nancy Linde. Produced by WGBH Television of Boston, Massachusetts.

The Royal Society. 1998. *Genetically Modified Plants for Food Use.* Ref: 2/98, London: The Royal Society.

———. 2002. *Genetically Modified Plants for Food Use and Human Health— An Update.* Policy Document 4/02, London: The Royal Society.

———. 2009. *Reaping the Benefits: Science and the Sustainable Intensification of Global Agriculture.* RS Policy Document 11/09, London: The Royal Society.

The Scottish Government. 2015. *GM Crop Ban.* August 9. Accessed January 8, 2016. http://news.scotland.gov.uk/News/GM-crop-ban-1bd2.aspx.

The Wall Street Journal. 2010. "Ag Department Uproots Science." December 27. Accessed February 4, 2016. http://www.wsj.com/articles/SB1000142405274870358120457600 33611631362824.

Then, Christoph. 2010a. "Agro-Biotechnology: Testbiotech Opinion on EFSA's Draft Guidance on the Environmental Risk Assessment of Genetically Modified Plants." Testbiotech, Munich.

————. 2010b. "Testbiotech Analysis of EFSA Guidance on the Environmental Risk Assessment of Genetically Modified Plants." TestBiotech. Accessed February 17, 2013. http://testbiotech.org/en/node/712.

Then, Christoph, and Andreas Bauer-Panskus. 2010. "European Food Safety Authority: A Playing Field for the Biotech Industry." Accessed February 17, 2013. http://www.testbiotech.org/en/node/431.

Toke, Dave. 2004. *The Politics of GM Food: A Comparative Study of the UK, USA, and EU.* New York: Routledge.

Tudge, Colin. 2003. "Where to Draw the Line." *New Scientist* 178, no. 2395: 23.

UK Parliament. 2015. *Science and Technology Committee—Role.* November 9. Accessed November 9, 2015. http://www.parliament.uk/business/committees/committees-a-z/commons-select/science-and-technology-committee/role/.

UNESCO. 1992. "The Rio Declaration on Environment and Development." *UNESCO.* Accessed May 7, 2013. http://www.unesco.org/education/nfsunesco/pdf/RIO_E.PDF.

Urry, John. 1992. "Reviewed work(s): Post-Marxist Alternatives. The Construction of Social Orders by Nicos Mouzelis." *The British Journal of Sociology* 43, no. 4: 692–93.

Warren, Mark. 1981. "Nietzsche's Concept of Ideology." *Theory and Society* 13, no. 4: 541–65.

Weale, Albert. 2007. "The Precautionary Principle in Environmental Policies." In *The Sage Handbook of Environment and Society,* edited by Jules Pretty et al., 590–600. London: Sage Publications.

Webster, Ben. 2015. "Britain Must be Free to Grow GM Food, Says Minister." January 8. Accessed November 7, 2015. http://www.thetimes.co.uk/tto/environment/article4316881.ece.

Welsh, Rick, and David Ervin. 2006. "Precaution as an Approach to Technology Development: The Case of Transgenic Crops." *Science, Technology, and Human Values* 31, no. 2: 153–72.

White, Lynn Jr. 1964. *Medieval Technology and Social Change.* Oxford: Oxford University Press.

WHO. 2016. *Micronutrient Deficiencies: Vitamin A Deficiency.* January 1. Accessed January 14, 2016. http://www.who.int/nutrition/topics/vad/en/.

Wilsdon, James and Rebecca Willis. 2004. *See-through Science.* London: Demos.

Wilsdon, James, Brian Wynne and Jack Stilgoe. 2005. "The Public Value of Science: Or How to Ensure that Science Really Matters." London: Demos.

Wilson, Allison K., Jonathan R. Latham and Ricarda Steinbrecher. 2006. "Transformation-induced Mutations in Transgenic Plants: Analysis and Biosafety Implications." *Biotechnology and Genetic Engineering Reviews* 23: 209–34.

Winner, Langdon. (1977). *Autonomous Technology: Technics-out-of-control as a Theme in Political Thought.* Cambridge, MA: The MIT Press.

Woolgar, Steve and Javier Lezaun. 2013 "The Wrong Bin Bag: A Turn to Ontology in Science and Technology Studies?" *Social Studies of Science* 43, no. 3: 321–40.

World Health Organization. 2005. "Modern Food Biotechnology, Human Health and Development: An Evidence-based Study." *Biotechnology.* Food Safety Department. Accessed December 5, 2012. http://www.who.int/foodsafety/publications/biotech/biotech_en.pdf.

World Trade Organization. 2008. Dispute Settlement: Dispute *DS291.* Accessed December 5, 2012. http://www.wto.org/english/tratop_e/dispu_e/cases_e/ds291_e.htm.

Wyatt, Sally. 2008. "Technological Determinism is Dead; Long Live Technological Determinism." In *The Handbook of Science and Technology Studies,* edited by Edward

J. Hackett, Olga Amsterdamska, Michael Lynch and Judy Wajcman, 165–80. London: The MIT Press.

Wynne, Brian. 1992 "Uncertainty and Environmental Learning: Reconceiving Science and Policy in the Preventive Paradigm." *Global Environmental Change* 2, no. 2: 111–27.

———. 2010a. "GeneWatch PR: New GM Dialogue Resignation Welcomed." *GeneWatch UK.* Accessed December 4, 2012. http://www.genewatch.org/uploads/f03c6d-66a9b354535738483c1c3d49e4/Resignationletter31May10.doc.

———. 2010b. "Brian Wynne Resigns from FSA Group Over GM: Letter of Resignation." *Lancaster University: Research Directory.* June 2. Accessed December 4, 2012. http://www.research.lancs.ac.uk/portal/en/clippings/brian-wynne-resigns-from-fsa-group-over-gm%28f96d22dc-def3-4d20-b043-7a91980563c8%29.html.

Yearley, Steven. 2005. *Making Sense of Science: Understanding the Social Study of Science.* London: Sage Publications.

Young, Gary. 1976. "The Fundamental Contradiction of Capitalist Production." *Philosophy and Public Affairs* 5, no. 2: 196–234.

Zahabi-Bekdash, Lara El and James V. Lavery. 2010. "Achieving Precaution through Effective Community Engagement in Research with Genetically Modified Mosquitoes." *Asia Pacific Journal of Molecular Biology and Biotechnology* 18, no. 2: 247–50.

Zapfel, Peter. 2002. "Science and Economics in Climate Change and Other International Negotiations." In *Transboundary Environmental Negotiation: New Approaches to Global Cooperation*, edited by Lawrence Susskind, William Moomaw and Kevin Gallagher, 130–53. San Francisco: Jossey-Bass.

INDEX

www.ingramcontent.com/pod-product-compliance
Lightning Source LLC
Chambersburg PA
CBHW022350280326
41935CB00007B/144